Distinguished Dissertations

Springer
London
Berlin
Heidelberg
New York
Barcelona
Hong Kong
Milan
Paris
Singapore
Tokyo

Other titles published in this Series:

Search and Planning Under Incomplete Information: A Study Using Bridge Card Play
Ian Frank

Theorem Proving with the Real Numbers
John Harrison

Hardware Evolution: Automatic Design of Electronic Circuits in Reconfigurable Hardware by Artificial Evolution
Adrian Thompson

User-Developer Coooperation in Software Development
Eamonn O'Neil

A Combination of Geometry Theorem Proving and Nonstandard Analysis, with Application to Newton's Principia
Jacques Fleuriot

Accurate Visual Metrology from Single and Multiple Uncalibrated Images
Antonio Criminisi

Asynchronous System-on-Chip Interconnect
John Bainbridge

John MacCormick

Stochastic Algorithms for Visual Tracking

Probabilistic Modelling and Stochastic Algorithms for Visual Localisation and Tracking

Springer

John MacCormick
Systems Research Center, Compaq Computer Corporation, 130 Lytton Avenue, Palo Alto, CA 94301-1044, USA

Series Editor
Professor C.J. van Rijsbergen
Department of Computing Science, University of Glasgow, G12 8RZ, UK

British Library Cataloguing in Publication Data
MacCormick, John
Stochastic algorithms for visual tracking. - (Distinguished dissertations)
1.Computer vision - Mathematical models 2. Algorithms
3.Stochastic processes
I.Title
006.3'7
ISBN-13: 978-1-4471-1176-4 e-ISBN-13: 978-1-4471-0679-1
DOI: 10.1007/978-1-4471-0679-1

Library of Congress Cataloging-in-Publication Data
A catalog record for this book is available from the Library of Congress.

Distinguished Dissertations ISSN 1439-9768

ISBN-13: 978-1-4471-1176-4 Springer-Verlag London Berlin Heidelberg
a member of BertelsmannSpringer Science+Business Media GmbH
http://www.springer.co.uk

Softcover reprint of the hardcover 1st edition 2002

Typesetting: Camera-ready by author

34/3830-543210 Printed on acid-free paper SPIN 10868264

Preface

Computer vision is hard. One way to make progress on hard problems is to develop a coherent mathematical theory which describes them. This book develops one such theory for the field of visual tracking, using probabilistic models and stochastic algorithms. In an applied science like computer vision, however, theory is of little use without practice. Hence, the book also demonstrates the use of this theory in real visual tracking systems.

The stochastic algorithms employed here — variously known as "particle filters", or the "Condensation algorithm" — are introduced in Chapter 2 via a formulation based on sequences of weighted particles. The properties of the particle filter are proved rigorously, yet presented in a manner intended to be accessible to a general computer vision audience. Chapter 3 introduces the other essential ingredient for visual tracking: a set of probabilistic models for object localisation. These models, called "contour likelihoods", are based on the statistical properties of image features. Chapter 4 explains the use of contour likelihoods in localisation and tracking problems, and Chapter 5 uses Markov random fields to generalise the contour likelihood for modelling the occlusion of foreground objects. Contour likelihoods can also be derived for collections of more than one object. Chapter 6 shows how such derivations lead to an "exclusion principle" which enables the successful tracking of foreground objects which occlude each other. Finally, Chapter 7 introduces the technique of "partitioned sampling", which dramatically reduces the computational expense of particle filters in certain problems. As in the previous chapters, I have tried to prove the results in a rigorous yet accessible manner. Readers equipped with a basic knowledge of mathematics, engineering, or computer science will have no difficulty in following the text.

This book arose from my Ph.D. thesis at the University of Oxford. My greatest debt is to Andrew Blake, who was my Ph.D. supervisor; he provided many ideas and much insight. I am also indebted to my colleagues at the Oxford Visual Dynamics Laboratory during the late 1990s: Benedicte Bascle, Colin Davidson, Jonathan Deutscher, Michael Isard, Bob Kaucic, Ben North, Jens Rittscher, Josephine Sullivan, and Andrew Wildenberg. Michael deserves a special vote of thanks because he wrote a lot of the code that I used and was also a very good friend. Many other people helped to keep me sane or were nice to me even when they realised they were failing in this goal; among them are Bear and Christie Bolton, Paul Harvey and his family, Andy Marsham, Jon Scammon, and everyone from OUHC.

For financial support, I am very grateful to the trustees of the Eliot Davis Memorial Scholarship, the trustees of the Jowett Scholarship, and the European Union IMPROOFS grant. Three Oxford colleges — Exeter, Balliol, and Linacre — have assisted me, and I am grateful to each for its generosity. The Systems Research Center of Compaq Computer Corporation supported me during the preparation of the manuscript, and the British Computer Society enabled its publication.

My family all live far away but are close in other ways. In fact they probably helped out much more than they realised, and I would especially like to thank Mum, Dad, Andrew, Jude and G&G, P&B, JGJS&A, RLVSK&L. The book is dedicated to my parents, with love and respect.

John MacCormick
Palo Alto, California
January, 2002

Contents

1

Introduction and background

Two of the important problems in computer vision are

- to locate, and determine the configuration of, a known object in a static image
- to locate, and determine the configuration of, a (moving, deforming) known object in each frame of a video sequence

The word "object" here could refer to a given physical object (the coffee mug in front of me), or to any element of a class of objects (apples, or cars). No general algorithm for solving these problems by computer is known. Indeed, some would claim that no such algorithm exists. To solve these problems satisfactorily, the "known object" must be defined, and in at least some cases this requires the computer to *understand* a general notion (e.g. "cars") involving language. This in turn leads to a whole host of problems associated with language and consciousness which have been studied by philosophers, computer scientists and others working in artificial intelligence (e.g. Boden, 1990). Those who do not believe the human brain is equivalent to a Turing machine can plausibly appeal to our non-Turing-ness for an explanation as to why humans can solve the problems above with such apparent ease while computers perform so poorly. The rest of us must wait, think, and try to come up with better algorithms.

This book comprises a contribution to the waiting and thinking, while acknowledging that much more must be done before our computer vision algorithms can fit into the layman's idea of the phrase "artificial intelligence". A particular emphasis of the approach in this book is that statistics and probability underpin the algorithms and can be used to interpret the results. Each of the methods of localisation and tracking described later produces as output a probability distribution, which can be interpreted as the computer's "belief" about the state of the world.

To summarise: we would like to use a computer to make inferences about the configurations of objects in images and video sequences. There are two fundamental contributions to this objective in the book. The first is a coherent set of models which can be used to infer the location and configuration of objects based on simple properties (essentially their shape and colour), both in static images and as the objects move in video streams. These models, termed *contour likelihoods*, lead to statistical quantities such as likelihood ratios which can be easily interpreted. Refinements and extensions of the basic contour likelihood approach lead to models which are valid for partially occluded objects (the *Markov likelihood*), and to likelihoods which can be used for tracking more than one object simultaneously; the latter is an example of a "probabilistic exclusion principle".

The second contribution is to the theory of particle filters, and their use in computer vision applications. The theory and basic results on particle filters are collected and proved in a new, self-contained formulation based on *particle sets*. The exposition is rigorous yet accessible to the vision community, and methods of assessing the efficacy of particle filters are also introduced. Finally, a technique called *partitioned sampling* is described. Partitioned sampling dramatically improves the performance of an important class of particle filters, and thereby provides a solution to certain tracking problems that were previously unassailable by particle filtering methods.

1.1 Overview

The diverse nature of the problems to be tackled means that a unified literature survey would tend to confuse rather than elucidate. Instead, literature surveys are included in each chapter where they are needed. In particular, Sections 2.7 and 2.8 are surveys on aspects of particle filtering, Section 4.1 surveys object localisation, Section 6.1 surveys multiple-target tracking and Chapter 3 begins with a brief overview of generative models.

The content of the book is as follows. The next section of this chapter provides an introduction to *active contours* — the particular set of models used to describe and make inferences on images throughout the book.

Chapter 2 ("The Condensation algorithm") is a self-contained description of particle filters, which are known in the computer vision literature as the Condensation algorithm. Condensation is the algorithm used for all the tracking problems considered in the book, so Chapter 2 provides a proof of its asymptotic correctness and introduces concepts useful for assessing the effectiveness of particle filters.

The huge amount of data in images means that vision algorithms often consider only a much smaller set of data called *features*, such as edges or colour blobs, which are derived from the image in a prescribed manner. Statistical interpretation of the features then requires a *generative model* of the features — a statistical model describing how the features arise in typical images. Chapter 3 ("Contour likelihoods") is a description of the generative models used in this book, and of the standard types of statistical inference that can be performed on these models.

Chapter 4 is a description of how to apply the contour likelihoods of Chapter 3 to real localisation and tracking problems. The methods are demonstrated on a variety of static localisation problems, and real-time tracking tasks.

The generative models of Chapter 3 assume that certain types of image features are statistically independent; this is an approximation that is violated badly when an object is partly occluded. Chapter 5 ("Modelling occlusions using the Markov likelihood") is a self-contained solution to this problem which modifies the likelihoods by using a Markov random field.

Chapter 6 ("A probabilistic exclusion principle for multiple objects") generalises the previous contour likelihoods so that they can be used with more than one object. It turns out that the correct way of doing this leads to a probabilistic exclusion principle: the computer's hypotheses about the object configurations are prevented from coalescing unduly onto the same state.

As the final substantive chapter of the book, Chapter 7 ("Partitioned sampling") introduces a method for dramatically improving the effectiveness of certain particle filters which must operate in high-dimensional spaces. Partitioned sampling is particularly suited to tracking multiple objects and articulated objects, and examples of both are demonstrated. A particular highlight is a real-time hand tracking system of sufficiently good quality that it can be used as the input for a drawing package.[1]

Chapter 8, after some brief speculation on what might come next, directs the reader back here for a summary.

Some of the material in the book has appeared elsewhere, including in Blake et al. (1998), MacCormick and Blake (1998a, 1998b, 1999), MacCormick and Isard (2000), and Isard and MacCormick (2000). Video sequences and other material related to the book are available from the following web site:

`http://www.robots.ox.ac.uk/~jmac/visualtracking`

1.2 Active contours for visual tracking

The human visual system provides empirical proof that accurate tracking of moving, deforming objects is an information-processing problem with a robust, real-time solution. A full understanding of this (human) solution is still well beyond our grasp, and current solutions using *artificial* intelligence must make simplifying assumptions and accept less complete results to make progress. Several different paradigms for these simplifying assumptions are popular in the literature. If we accept that complete reconstruction of the "plenoptic function" (Adelson and Bergen, 1991) — or even more generally, a complete reconstruction of both the 3D structure of the world and its illuminance properties — is not a realistic goal, then we must instead adopt some basis for segmenting the world and our images of it into meaningful blocks. A simplistic taxonomy of approaches to this segmentation task might list the following: "layers" (e.g. Wang and Adelson, 1993; Baker et al., 1998), in which the world is taken to

[1] This was developed in collaboration with Michael Isard.

consist of cardboard cutouts; "texture" (e.g. Chellappa and Chatterjee, 1985; Malik et al., 1999), in which the world consists of objects defined by homogeneous textures; and "contours" or "snakes" (e.g. Kass et al., 1987; Blake and Isard, 1998; Terzopoulos and Szeliski, 1992), in which the world consists of objects defined by boundaries with given properties.

We have to start somewhere to get somewhere else. So this book adopts the contour approach, and builds on a theory of shape and motion espoused by Andrew Blake and his co-workers at the University of Oxford. Termed "active contours", this approach is described in detail in a book of the same name by Blake and Isard (1998). This section of the book will cover the bare bones of this theory (splines, shape spaces, and dynamical models using auto-regressive processes) for completeness, but the interested reader is urged to consult the *Active Contours* book, as well as research papers such as those by Blake et al. (1993, 1995), Curwen and Blake (1992), and Reynard et al. (1996) and, for a broader perspective, the *Active Vision* collection (Blake and Yuille, 1992). Important related work includes the Active Shape Models of Cootes, Taylor, and others in Manchester (e.g. Cootes et al., 1995), and the snakes of Kass et al. (1987). Some detailed applications of the active contour approach are given in Oxford Ph.D. theses by Wildenberg (1997), Kaucic (1997), Isard (1998), and North (1998).

1.2.1 Splines and shape space

Each tracking or localisation problem addressed by this book involves a distinguished class of objects called "targets" or "target objects". Examples of typical classes of targets include cars, bicycles, humans, and hands. A given target object is described by its outline, which is modelled as a B-spline (Bartels et al., 1987). B-splines are a convenient way of representing smooth, natural-looking shapes by specifying only a small number of "control points". Specifically, if the coordinates of the control points are $(x_1, y_1), \ldots (x_n, y_n)$ then the B-spline is a curve $(x(s), y(s))^{\mathrm{T}}$ parameterised by a real variable s on an interval of the real line:

$$\begin{pmatrix} x(s) \\ y(s) \end{pmatrix} = B(s) \begin{pmatrix} \vec{x} \\ \vec{y} \end{pmatrix}, \tag{1.1}$$

where $B(s)$ is a $2 \times 2n$ matrix whose entries are polynomials in s, and $\vec{x}$, $\vec{y}$ are $n \times 1$ column vectors containing the x- and y-coordinates of the control points respectively. The key point to note is that for a given B, the space of all contours is finite-dimensional: indeed it is isomorphic to $\mathbb{R}^{2n}$, and can be given an inner product which makes the concept of "distance" between two splines meaningful (see Appendix A.1).

We will call any such B-spline a *contour*. Figure 1.1 gives an example of a mouse-shaped contour with seven control points. Typical objects require 5–20 control points to produce a contour that, for a human observer, appears to match the target object very closely. The vector space of such contours has 10–40 dimensions, which is often undesirably large. Hence we generally work in a vector subspace of $\mathbb{R}^{2n}$ termed the *shape space* and denoted by $\mathcal{X}$.

An element $\mathbf{x} \in \mathcal{X}$ is related to the control point coordinates $\vec{x}, \vec{y}$ by a linear transformation with a fixed offset:

$$\begin{pmatrix} \vec{x} \\ \vec{y} \end{pmatrix} = W\mathbf{x} + Q. \tag{1.2}$$

If the shape space has d dimensions, then W is the $2n \times d$ *shape matrix*, $\mathbf{x}$ is a $d \times 1$ column vector generally just referred to as the *configuration* or *state*, and Q is a $d \times 1$ vector called the *template* for this object.

Given a template for a rigid planar object, it is easy to define a shape space $\mathcal{X}$ corresponding to either translations, 2D Euclidean similarities or affine transformations[2] of the template. If the target object is articulated, non-rigid, or non-planar, it is still often possible to work in a linear space of low dimension by extending $\mathcal{X}$. This can be done on theoretical grounds, or by using key-frames from training data followed by principal components analysis (Jollike, 1986; Cootes et al., 1995). Actually, none of the object localisation techniques in this book require $\mathcal{X}$ to be a vector space. $\mathcal{X}$ can be any sub-manifold of the spline space and discussions will treat it as such.[3] Even the tracking algorithms discussed later do not really require a linear shape space, although linearity is required to implement the dynamical models discussed in the next section.

1.2.2 Dynamical models using auto-regressive processes

Any tracking or filtering algorithm requires a model of how the system is expected to evolve over time. The computer vision community seems to divide into two camps here. On the one hand, one can attempt explicit physical realism by modelling kinematics and other forces (e.g. Robertson and Schwertassek, 1988; Bregler and Malik, 1998). The other chief approach is the *auto-regressive process* or ARP (Lutkepohl, 1993), which has been used in many speech and vision applications (e.g. Bobick and Wilson, 1995; Rabiner and Bing-Hwang, 1993; Black and Yacoob, 1995). This book adopts second-order ARPs exclusively. It has been our experience that these models capture a rich variety of motions of interest in computer vision tracking problems, while being both straightforward to learn from training data and easy to implement efficiently in real-time algorithms.

A second-order ARP expresses the state $\mathbf{x}_t$ at time t as a linear combination of the previous two states and some Gaussian noise:

$$\mathbf{x}_t = A_2\mathbf{x}_{t-2} + A_1\mathbf{x}_{t-1} + Bw,$$

where A_1, A_2, B are fixed $d \times d$ matrices and w is a $d \times 1$ vector of i.i.d. standard normal variates. Blake and Isard (1998) give some useful techniques for setting A_1, A_2, B by physical reasoning for relatively complex tracking problems, and

[2] For planar objects, these are equivalent to 3D Euclidean transformations with perspective effects neglected.

[3] Authors using contours seem to have restricted themselves to linear configuration spaces because of various computational advantages and because the Kalman filter requires a linear state space.

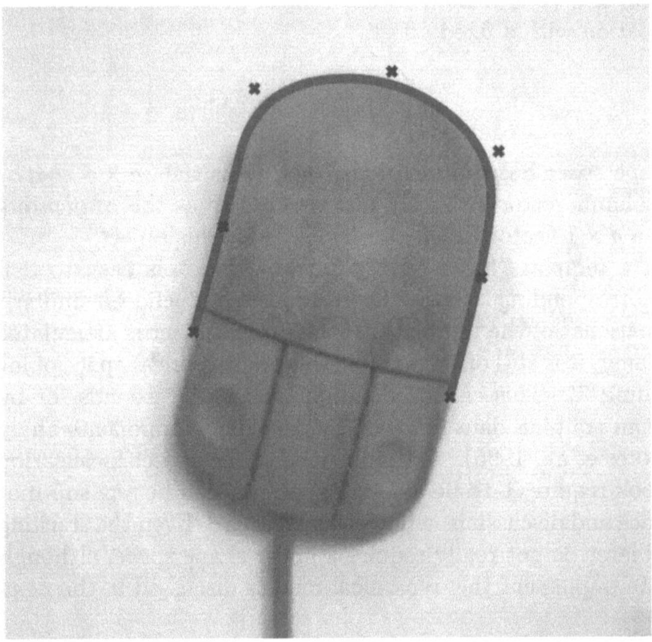

Figure 1.1: ***A contour.*** *Throughout the book, a* contour *is a B-spline like this one, specified by control points shown here as black crosses. Note that the contour does not interpolate the control points.*

then bootstrapping by taking maximum likelihood estimates based on longer and longer sequences of successful tracking. This is the approach used for all tracking in this book. The expectation-maximisation algorithm has also been used for learning ARPs; two computer vision applications can be found in North (1998), and Pavlovic et al. (1999).

1.2.3 Measurement methodology

To perform inferences on real data, one must not only acquire the data, but also possess a model of the data acquisition process. This section describes the process itself, while Chapter 3 addresses ways of modelling it. A plethora of tools have been used for this by the image analysis community, and we choose one based on its utility for tracking moving and deforming objects.

The method is demonstrated in Figure 1.2. At fixed[4] points along the B-spline, line segments normal to the contour are cast onto the image. These normals are called *measurement lines*; typically they will be spaced equally around the contour, with between one and five measurement points per span of

[4] To be precise, the measurement points are not fixed on the contour at all — this concept makes no sense when the contour's shape can vary. The measurement points are actually placed at fixed values of the B-spline parameter s in equation 1.1.

the B-spline. Often the length of the measurement line is fixed in advance, and it is almost always assumed that the measurement line is symmetric about the measurement point. Next, a one-dimensional feature detector (a simple edge detector, for instance) is applied to the image intensity along each measurement line. For each line, the result is a list of numbers describing the distance from each feature found to the measurement point; these numbers are called the *feature innovations.*[5]

Chapter 3 describes some detailed models which can be used to perform inference on the distribution of the features which are detected by this method. Section 3.2.3 describes a different method of measuring the image, in which the measurement lines are fixed *in the image* rather than being attached to the contour.

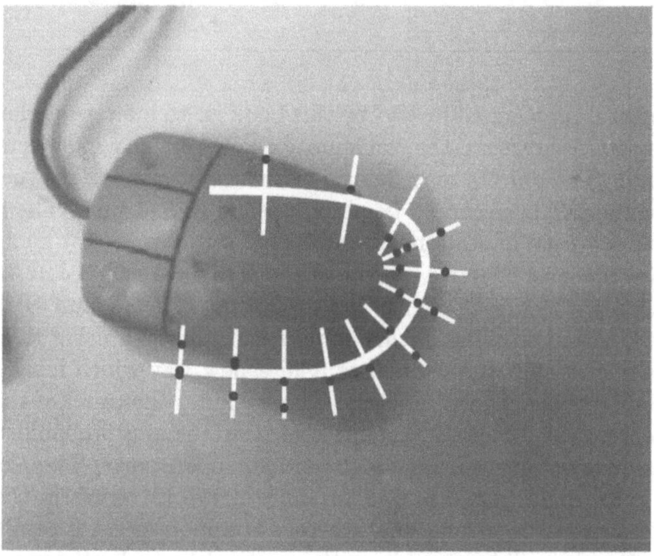

Figure 1.2: ***Measurement methodology and notation.*** *The thick white line is* $\mathbf{x}$ *— a mouse-shaped contour in some hypothesised configuration. The thin lines are measurement lines, along which a one-dimensional feature detector is applied. Black dots show the output of the feature detector. The distance of a feature from the contour is called its* innovation.

[5]The contour is assumed to have an orientation, and a negative feature innovation indicates the feature is in the interior of the contour. (By convention, the orientation is chosen so that the interior is on the left of an ant traversing the contour in the positive direction.) Note that both open and closed contours have an interior according to this definition.

2

The Condensation algorithm

A fundamental problem addressed by this book is: to infer, from data captured with a single video camera, the configurations of some known objects, as a function of time. The objects may deform or move throughout the sequence, and if possible the inference should be accomplished in real time. The background is not assumed rigid or stationary, and its 3D structure is not known.

Perhaps the most effective tool for this task to have emerged in the computer vision literature is the Condensation algorithm of Isard and Blake (1998a). Most of the visual tracking in this book is accomplished using this algorithm; this chapter describes and formalises it in a new way which makes the power and flexibility of the algorithm clear. This includes the proof of the "Condensation theorem" which shows that the algorithm remains probabilistically valid under various combinations of its constituent operations. The Condensation theorem is an important contribution since previous proofs in the computer vision literature were incomplete, whereas the more general results from the statistics literature did not address the particular stochastic operations used in the contour tracking framework. For the first time, a rigorous proof of the correctness of Condensation is made accessible to the computer vision community. Another innovation is the introduction of Condensation diagrams,[1] which enable the myriad possible implementations of Condensation to be described both accurately and concisely.

The final two sections of the chapter are mini-appendices, describing respectively the history of, and some alternatives to, the Condensation algorithm.

2.1 The basic idea

Throughout it is assumed that there are T frames of data to be processed, but that at time t only data from times $1, \dots t-1$ are available, so that we are in a filtering rather than smoothing framework. The measurements acquired from

[1] For this I am indebted to Michael Isard, who helped develop the notation and wrote the LaTeX package which displays them.

frame t are labelled $\mathbf{Z}_t$, and will generally consist of a list of edgels observed in the image, as discussed in Section 1.2.3 (although of course the measurements could actually be any other function of the image, including the whole image itself[2]). The configuration of the target objects at time t is $\mathbf{x}_t$, and the measurements $\mathbf{Z}_t, t = 1, \ldots T$ are assumed to be conditionally independent, given the $\mathbf{x}_t, t = 1, \ldots T$. The measurements acquired up to frame t are denoted $\mathcal{Z}_t$:

$$\mathcal{Z}_t = \{\mathbf{Z}_1, \ldots \mathbf{Z}_t\}.$$

The objective of filtering is to apply Bayes' theorem at each time-step, obtaining a posterior $p_t(\mathbf{x}_t|\mathcal{Z}_t)$ based on all available information:

$$p_t(\mathbf{x}_t|\mathcal{Z}_t) = \frac{p_t(\mathbf{Z}_t|\mathbf{x}_t)p_{t-1}(\mathbf{x}_t|\mathcal{Z}_{t-1})}{p_t(\mathbf{Z}_t)}, \tag{2.1}$$

where we can write $p_t(\mathbf{Z}_t|\mathbf{x}_t)$ instead of $p_t(\mathbf{Z}_t|\mathbf{x}_t, \mathcal{Z}_{t-1})$ because of the conditional independence assumption. As usual in filtering theory, a model for the expected motion between time-steps is adopted. This takes the form of a conditional probability distribution $p_t(\mathbf{x}_t|\mathbf{x}_{t-1})$ termed the *dynamics*. Using the dynamics, equation 2.1 can be re-written as

$$p_t(\mathbf{x}_t|\mathcal{Z}_t) = \frac{p_t(\mathbf{Z}_t|\mathbf{x}_t)\int_{\mathbf{x}_{t-1}} p_t(\mathbf{x}_t|\mathbf{x}_{t-1})p_{t-1}(\mathbf{x}_{t-1}|\mathcal{Z}_{t-1})\,d\mathbf{x}_{t-1}}{p_t(\mathbf{Z}_t)}. \tag{2.2}$$

This is the equation which a filter must calculate or approximate. In the particularly well-known and easily implemented case of a Kalman filter (Gelb, 1974), the observation density $p_t(\mathbf{Z}_t|\mathbf{x}_t)$ is assumed Gaussian, and the dynamics are assumed linear with additive Gaussian noise. Unfortunately, it is an empirical fact that the observation densities occurring in visual tracking problems are not at all Gaussian; this was the original motivation for Isard and Blake to introduce the Condensation algorithm. The fundamental idea behind Condensation is simply to simulate equation 2.2. First note that equation 2.2 can be re-written as a sequence of two operations on density functions:

$$\begin{array}{ccccc} p_{t-1}(\mathbf{x}_{t-1}|\mathcal{Z}_{t-1}) & \longmapsto & p_{t-1}(\mathbf{x}_t|\mathcal{Z}_{t-1}) & \longmapsto & p_t(\mathbf{x}_t|\mathcal{Z}_t), \\ & \text{convolve with} & & \text{multiply by} & \\ & \text{dynamics} & & \text{observation} & \\ & p(\mathbf{x}'|\mathbf{x}) & & \text{density } p(\mathbf{Z}|\mathbf{x}') & \end{array} \tag{2.3}$$

where the final division by $p_t(\mathbf{Z}_t)$ has been omitted — this normalisation step will be achieved automatically by the method adopted.

The simulation uses the idea of a *particle set*: a list of n pairs $(\mathbf{x}_i, \pi_i), i = 1, \ldots n$, where $\mathbf{x}_i \in \mathcal{X}$ (the configuration space) and $\pi_i \in [0, 1]$ with $\sum_{i=1}^{n} \pi_i =$

[2]Isard and Blake (1998a) allow $\mathbf{Z}$ to be the entire image, and offer arguments as to why their observation density depending only on edgels can be regarded as an approximation to the true density depending on all grey-levels. I prefer to regard the measurement process as a black box, which outputs *features* $\mathbf{Z}_t$. Rigorous statistical inference proceeds on the basis of the black box output, not the original grey-levels or RGB values.

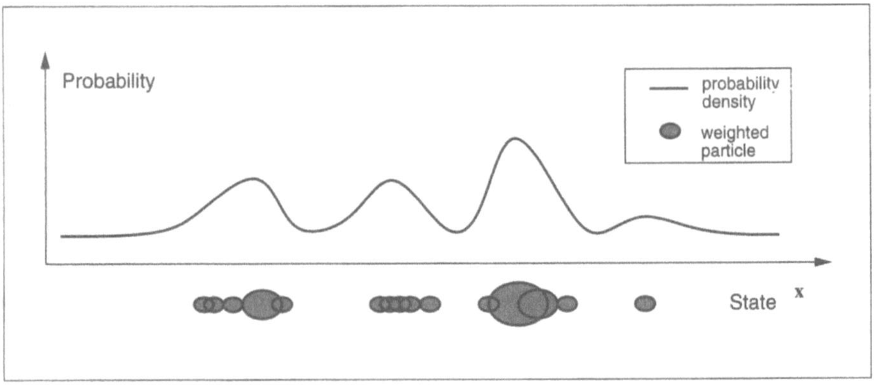

Figure 2.1: ***Particle sets.*** *A particle set is an approximate representation of a probability distribution: picking one of the grey ellipses with probability proportional to its area is approximately the same as drawing randomly from the continuous probability density function shown. (Figure reproduced from (Isard and Blake, 1998a).)*

1. The particle set is meant to represent a probability distribution $p(\mathbf{x})$, in the sense that choosing one of the $\mathbf{x}_i$ with probability π_i should be approximately the same as drawing a random sample from the distribution $p(\mathbf{x})$ (Figure 2.1). To simulate equation 2.2, all we need is a way of performing the operations "convolve with dynamics" and "multiply by observation density" on particle sets. This turns out to be very easy. To convolve with dynamics $p(\mathbf{x}'|\mathbf{x})$, just replace each particle $\mathbf{x}_i$ with a random draw from $p(\mathbf{x}'|\mathbf{x}_i)$. To multiply by the observation density $p(\mathbf{Z}|\mathbf{x})$, just multiply each weight π_i by $p(\mathbf{Z}|\mathbf{x}_i)$, and normalise so that the new weights sum to unity. It turns out that another operation termed *resampling* is required too; the reason for this is explained in the next section. The resampling operation consists of choosing n of the particles from the previous step by sampling with replacement; particles are selected with probabilities proportional to their weights. The whole process is summarised on a *Condensation diagram*:

$$\underset{(a)}{\boxed{\text{prior}}} \longrightarrow \boxed{\sim} \longrightarrow \underset{(b)}{\left\langle * p(\mathbf{x}'|\mathbf{x}) \right\rangle} \rightarrow \underset{(c)}{\left\langle \times p(\mathbf{Z}|\mathbf{x}') \right\rangle} \rightarrow \underset{(d)}{\boxed{\text{posterior}}} \tag{2.4}$$

where the $\sim$ symbol denotes resampling, $*$ denotes convolving with dynamics, and $\times$ denotes multiplication by the observation density. (The labels (a)–(d) correspond to the example given shortly.) These notions are formalised in Section 2.3, but first let's look at an example.

Example Figure 2.2 shows a simulation of the Condensation diagram 2.4. The configuration space $\mathcal{X}$ is two-dimensional, so the position $\mathbf{x} = (x_1, x_2)$ of each particle is represented by its position on the plane, and the weight π

of a particle is proportional to its area. Each panel (a)–(d) corresponds to a labelled position on the Condensation diagram 2.4. Panel (a) shows a particle set representing a Gaussian prior centred on the middle of the configuration space. The true position of the target is shown as a cross. In panel (b), the particle set has been resampled. The particles now have equal weights; there is the same number of particles as in (a), but the more heavily weighted particles have been replicated many times whereas some particles from (a) were not selected at all by the resampling process. Panel (c) shows the results of applying the dynamics (in this case additive Gaussian dynamics) to (b). The posterior (d) is created from (c) by multiplying the (equal) weights by the likelihood function, which is shown for reference in Figure 2.3. This likelihood function is a Gaussian centred on the true position of the target.

Those who have followed the evolution of Condensation in the vision literature might wonder at the need for a proof of the algorithm's correctness. In fact, there is such a need: the proof in Isard and Blake (1998a) contained a flaw, which I believe can only be rectified by the relatively heavy mathematical machinery to be invoked in the rest of this chapter. Recall that Condensation consists of iterating equation 2.4 once for every frame in the video sequence. Isard and Blake showed that equation 2.4 itself was valid, but they assumed the particle set representing the prior consisted of *independent* draws from the prior. They then used induction to deduce the validity for a finite number of iterations, but unfortunately the inductive hypothesis does not hold: after the resampling operation has been performed, the particles in a given particle set are not independent. Hence the need for a more general approach.

2.2 Formal definitions

Before we can prove anything useful about Condensation, some precise definitions are needed. Recall that the configuration space is denoted $\mathcal{X}$. Throughout the book, $\mathcal{X}$ is assumed to be a compact[3] subset of $\mathbb{R}^d$. Because we want to prove results about the asymptotic properties of Condensation as the number of particles n becomes large, we cannot work with a fixed value of n. We really need to consider the *sequence* of possible Condensation algorithms that would be produced by setting $n = 1, 2, \ldots$ in turn. The formal definition of a particle set captures this notion by considering a triangular array of random variables.

Definition (Particle set) A *particle* is a pair $(\mathbf{x}, \pi)$ where $\mathbf{x} \in \mathcal{X}$ and $\pi \in [0, 1]$; π is called the *weight* of the particle. A *particle set* S is a sequence of finite sets of random variables whose values are particles: the nth member of the sequence is a set of n random variables

$$S_n = \{(\mathbf{X}_1^{(n)}, \Pi_1^{(n)}), \ldots (\mathbf{X}_n^{(n)}, \Pi_n^{(n)})\},$$

such that $\sum_{i=1}^n \Pi_i^{(n)} = 1$. In other words, S is an infinite triangular array of

[3]That is, closed and bounded.

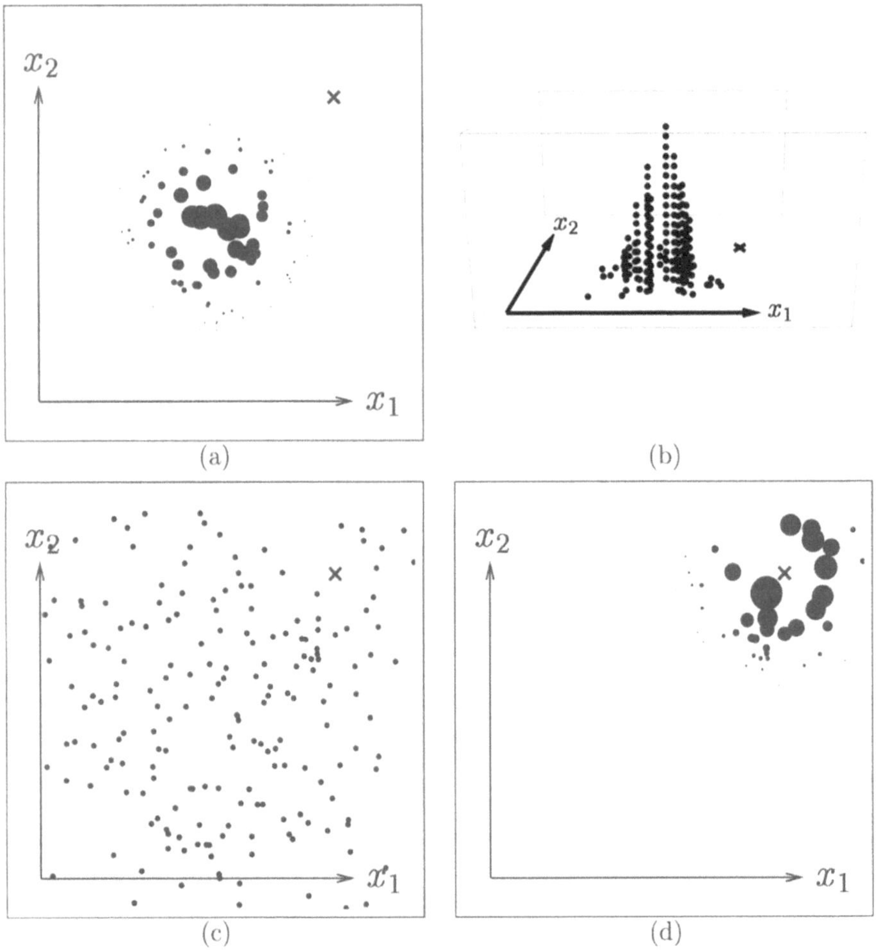

Figure 2.2: ***One time-step of the Condensation algorithm.*** *The true position of the target in this two-dimensional configuration space is shown as a cross; particles representing a probability distribution are shown as circles whose areas are proportional to their weights. Each step shown is one stage in the Condensation diagram of equation 2.4. (a) The prior. (b) After the resampling operation. Every particle now has the same weight, but many are replicated; these are shown stacked on top of each other. (c) After the dynamics have been applied. (d) After reweighting by the likelihood (see Figure 2.3).*

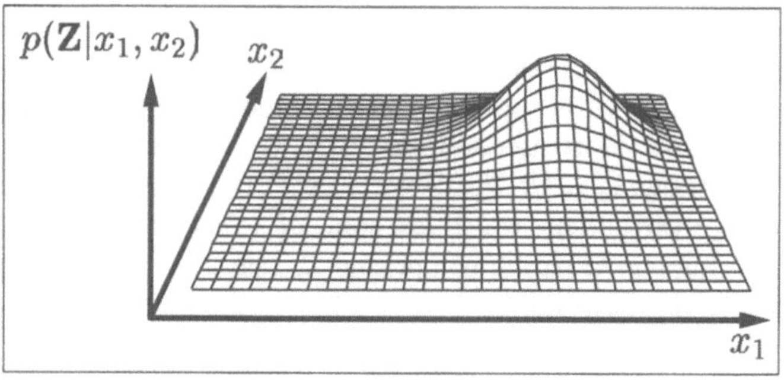

Figure 2.3: ***The likelihood used for reweighting in Figure 2.2.*** *In this simple example, the likelihood is a Gaussian centred on the true position of the target.*

particle-valued random variables, so $S = (S_1, S_2, \ldots)$ with

$$S_1 = \{(\mathbf{X}_1^{(1)}, \Pi_1^{(1)})\},$$
$$S_2 = \{(\mathbf{X}_1^{(2)}, \Pi_1^{(2)}), (\mathbf{X}_2^{(2)}, \Pi_2^{(2)})\},$$
$$\vdots$$

Generally we blur the distinction between a random variable and its value, simply writing $(\mathbf{x}_i^{(n)}, \pi_i^{(n)})_{i=1}^n, n = 1, 2, \ldots$ for the particle set. Furthermore, the fact that $n = 1, 2, \ldots$ and the superscript n are often omitted; operations on particle sets will often be specified by their effect on the typical nth element $S_n = (\mathbf{x}_i, \pi_i)_{i=1}^n$.

Note that there is a terminological difficulty here: a "particle set" is *not* a set of particles — it is a sequence of sets whose elements are particle-valued random variables.

Example The random variables for a particle set are almost always specified by an *algorithm* for generating them, rather than explicitly writing down their density functions. For example, suppose $\mathcal{X}$ is the unit interval $[0, 1]$. Then the typical element of a particle set could be defined by specifying that for $i = 1, 2, \ldots n$, $\mathbf{x}_i^{(n)}$ is drawn independently from the uniform distribution on $[0, 1]$ and that $\pi_i^{(n)}$ is set to $1/n$ (note that $\pi_i^{(n)}$ *is* a random variable, but that its law assigns a probability mass of 1 to the value $1/n$). Writing Rect$[0, 1]$ for the uniform distribution on $[0, 1]$ this can be written

$$\mathbf{x}_i^{(n)} \vdash \text{Rect}[0, 1],$$
$$\pi_i^{(n)} = 1/n, \tag{2.5}$$

where the $\vdash$ symbol means $\mathbf{x}_i^{(n)}$ should be drawn randomly from the specified distribution, independently of all the other draws.

Of course, the only reason for introducing particle sets is that they can be used to approximate arbitrary probability distributions on the configuration space. The intuitive idea is that, when n is large, a random sample from the nth element of a particle set is approximately a random sample from a given probability distribution. More precisely, note that we can identify the nth element of a particle set, say $S_n = (\mathbf{X}_1^{(n)}, \Pi_1^{(n)}), \ldots (\mathbf{X}_n^{(n)}, \Pi_n^{(n)})$, with a stochastic recipe for generating distributions which are sums of δ-functions: a particular realisation of the random variables comprising S_n, say $(\mathbf{x}_1^{(n)}, \pi_1^{(n)}), \ldots (\mathbf{x}_n^{(n)}, \pi_n^{(n)})$, corresponds to the distribution

$$\sum_{i=1}^{n} \pi_i^{(n)} \delta_{\mathbf{x}_i^{(n)}}, \tag{2.6}$$

where $\delta_{\mathbf{x}}$ is a δ-function centred on $\mathbf{x}$. What we want to discuss is when such stochastic recipes for generating distributions converge to a single distribution $p(\mathbf{x})$: this is formalised by the notion of the distribution *represented* by a particle set (Section 2.2.2), but to do this properly we need some technical details.

2.2.1 Technical detail: convergence of distribution-valued distributions

Our objective here is to present a rigorous treatment of particle sets which is accessible to a general computer vision audience. To so do we must sacrifice some of the finer niceties of measure theory — for instance, we will talk of "the family of all probability measures on $\mathcal{X}$" without specifying a σ-algebra on $\mathcal{X}$, and will not mention when functions should be measurable. The reader who is equipped to handle such details is referred to Del Moral (1998), on whose ideas and notation we draw heavily for this section.

With such caveats out of the way, we can proceed with some definitions. Let $\mathcal{P}(\mathcal{X})$ be the space of all probability measures on $\mathcal{X}$. Let $\mathcal{P}(\mathcal{P}(\mathcal{X}))$ be the space of all probability measures on $\mathcal{P}(\mathcal{X})$. For a real-valued function $F : \mathcal{P}(\mathcal{X}) \to \mathbb{R}$, and $\Phi \in \mathcal{P}(\mathcal{P}(\mathcal{X}))$, we write

$$\Phi F = \int F(\mu)\Phi(d\mu). \tag{2.7}$$

This is an integration over all possible measures $\mu \in \mathcal{P}(\mathcal{X})$, and the notation may be unfamiliar to some readers. If confusion sets in, revert to the simple case in which Φ is a discrete measure on $\mathcal{P}(\mathcal{X})$ — say $\Phi = \sum_{i=1}^{n} w_i \delta_{p_i}$ for $p_1, p_2, \ldots p_n \in \mathcal{P}(\mathcal{X})$ and with $\sum_{i=1}^{n} w_i = 1$. Then the integration becomes a weighted sum, and $\Phi F = \sum_{i=1}^{n} F(p_i) w_i$.

Definition (Weak convergence of distribution-valued distributions) A sequence $\Phi_1, \Phi_2, \ldots$ in $\mathcal{P}(\mathcal{P}(\mathcal{X}))$ *converges weakly* to $\Phi \in \mathcal{P}(\mathcal{P}(\mathcal{X}))$ if for every continuous, bounded, real-valued function $F : \mathcal{P}(\mathcal{X}) \to \mathbb{R}$, we have

$$\lim_{n \to \infty} \Phi_n F = \Phi F.$$

This is the only definition we need to make particle sets rigorous, so it is worth noting!

2.2.2 The crucial definition: how a particle set represents a distribution

The next definition uses the concept of weak convergence of distribution-valued distributions from the previous section.

Definition (Representation of a probability distribution by a particle set) Let S be a particle set with configuration space $\mathcal{X}$, and let $p(\mathbf{x})$ be a probability distribution on $\mathcal{X}$ (if you like, $p \in \mathcal{P}(\mathcal{X})$). Following the discussion of equation 2.6 above, we identify the typical element S_n of S with a member of $\mathcal{P}(\mathcal{P}(\mathcal{X}))$ — S_n is a "distribution-valued distribution" since it is a stochastic recipe for choosing distributions in $\mathcal{P}(\mathcal{X})$; the distributions it chooses happen to be sums of δ-functions in the form of equation 2.6. Let $\mathbf{p}$ be the element of $\mathcal{P}(\mathcal{P}(\mathcal{X}))$ which assigns mass 1 to p and 0 to all other distributions (this notation will be used throughout the book). We say S *represents* $p(\mathbf{x})$ if

$$S_n \rightarrow \mathbf{p} \text{ as } n \rightarrow \infty. \tag{2.8}$$

Of course this is weak convergence in $\mathcal{P}(\mathcal{P}(\mathcal{X}))$ as defined in the last section. An equivalent statement to equation 2.8 is that for all bounded continuous $F : \mathcal{P}(\mathcal{X}) \rightarrow \mathbb{R}$,

$$\int F(\mu)S_n(d\mu) \rightarrow F(p) \text{ as } n \rightarrow \infty. \tag{2.9}$$

This follows from the definition of weak convergence above, because $\mathbf{p}F = F(p)$.

The most obvious example of a particle set representing a distribution is the "standard Monte Carlo" particle set.

Definition (Standard Monte Carlo particle set for $p(\mathbf{x})$) Let $p(\mathbf{x})$ be a distribution on $\mathcal{X}$. Then the *standard Monte Carlo particle set for* $p(\mathbf{x})$ is $S = (\mathbf{x}_i^{(n)}, \pi_i^{(n)})$ given by

$$\begin{aligned} \mathbf{x}_i^{(n)} &\vdash p(\mathbf{x}), \\ \pi_i^{(n)} &= 1/n, \end{aligned} \tag{2.10}$$

where, as above, $\vdash$ symbol means $\mathbf{x}_i^{(n)}$ should be drawn randomly from the specified distribution, independently of all the other draws. (Remember, we are deliberately not bothering to distinguish between random variables and the values they take: the $\mathbf{x}_i$ and π_i here are random variables, but any computer simulation of the particle set — such as Figure 2.4 — will pick actual values for them.)

In other words, S_n consists of n independent samples from the "target" distribution $p(\mathbf{x})$. We can't get far without knowing that the standard Monte Carlo set really does represent $p(\mathbf{x})$, which is what the next proposition tells us.

Proposition 2.1 (Standard Monte Carlo) *A distribution $p(\mathbf{x})$ is represented by its standard Monte Carlo particle set.*

Proof. This is Corollary A.3 of Appendix A.3. ■

Some readers may be surprised, after looking at the appendix, that the proof of this seemingly obvious result is so abstract. The reason is that we have proved a result much stronger than the one needed for Monte Carlo integration. To do Monte Carlo integration, we only need to know that $1/n \sum^n f(\mathbf{x}_i) \to \int f(\mathbf{x})\, dp(\mathbf{x})$ for a suitable class of functions f (and this follows immediately from the Law of Large Numbers, for most f and p). Proposition 2.1, on the other hand, asserts that our Monte Carlo method, considered as an algorithm for producing distributions in $\mathcal{P}(\mathcal{P}(\mathcal{X}))$, converges to the element $\mathbf{p} \in \mathcal{P}(\mathcal{P}(\mathcal{X}))$.

Examples The particle set described in the previous example (Section 2.2) is actually the standard Monte Carlo set for the uniform distribution on $[0,1]$, and hence it does represent this distribution. But there are many other particle sets which also represent the uniform distribution (Figure 2.4). For example,

$$\begin{aligned} \mathbf{x}_i^{(n)} &= i/n, \\ \pi_i^{(n)} &= 1/n, \end{aligned} \tag{2.11}$$

also represents the uniform distribution. We call this type of particle set "deterministic" because there is no randomness in the way it chooses the particle configurations or weights. The fact that this deterministic particle set represents the uniform distribution can actually be checked directly from the definitions; it boils down to the fact that the integral of any continuous function can be approximated as well as we please by step functions with equally-spaced steps. As a final example, let Tri$[0,1]$ be the "triangular" distribution on $[0,1]$ — so the pdf is $q(x) = 2x$ on $[0,1]$ and 0 elsewhere. Define a particle set by:

$$\begin{aligned} \mathbf{x}_i^{(n)} &\vdash \mathrm{Tri}[0,1], \\ \pi_i^{(n)} &\propto 1/\mathbf{x}_i^{(n)}. \end{aligned} \tag{2.12}$$

The constant of proportionality for the π_i is determined by the fact that $\sum_{i=1}^n \pi_i^{(n)} = 1$. After Section 2.3 we will return to prove that this too is (almost) a representation of the uniform distribution.[4] Figure 2.4 shows random samples from all three particle sets just described.

2.3 Operations on particle sets

This section introduces the three basic operations on particle sets used by Condensation: multiplication by a function, application of dynamics, and resampling. Each operation is first described in terms of the stochastic algorithm which implements it. Secondly, the effect of each operation on the distribution represented by a particle set is stated and proved.

[4] As we shall see, we should really restrict this distribution to intervals $[\delta, 1]$ for $\delta > 0$ — the fact that $q(0) = 0$ causes problems otherwise.

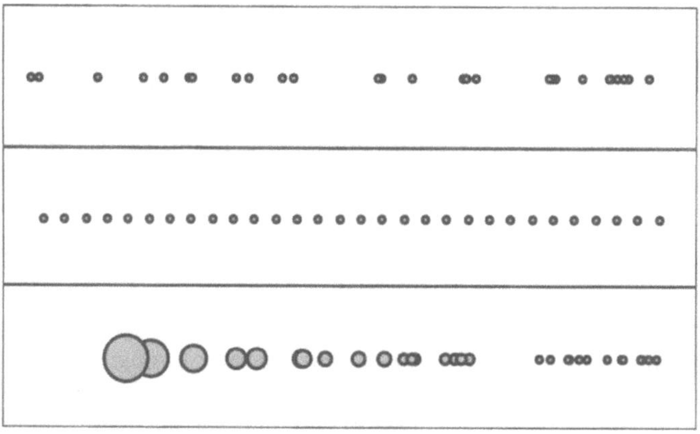

Figure 2.4: ***Particle sets representing the uniform distribution.*** *A sample from the element $n = 30$ is shown for each of three particle sets. Top: equation 2.5, which draws $\mathbf{x}_i^{(n)}$ from* Rect[0, 1]. *Middle: equation 2.11, which deterministically sets $\mathbf{x}_i^{(n)}$ to i/n. Bottom: equation 2.12, which draws $\mathbf{x}_i^{(n)}$ from the triangular distribution and sets $\pi_i^{(n)} \propto 1/\mathbf{x}_i^{(n)}$.*

2.3.1 Multiplication by a function

This is the simplest operation on particle sets, defined as follows.

Definition (Particle set multiplied by a function) Let S be a particle set on $\mathcal{X}$, with $S = \{(\mathbf{x}_i^{(n)}, \pi_i^{(n)})_{i=1}^n, n = 1, 2, \ldots\}$, and let $h(\mathbf{x})$ be a continuous, non-negative function on $\mathcal{X}$. (Note that because $\mathcal{X}$ is compact, h is therefore bounded and bounded away from zero.) Define a new particle set, $S' = \{(\mathbf{x}_i^{(n)'}, \pi_i^{(n)'})_{i=1}^n, n = 1, 2, \ldots\}$ by (the superscript n's are dropped):

$$\mathbf{x}'_i = \mathbf{x}_i$$
$$\pi'_i \propto h(\mathbf{x}_i)\pi_i.$$

We say S' is obtained from S by *multiplying* by h. If S represents a distribution $p(\mathbf{x})$, this is written on a Condensation diagram as

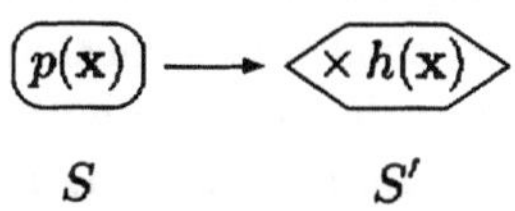

The effect of the multiplication operation on the distribution represented by a particle set is exactly what you would expect:

Theorem 2.2 *Let S, S' be as above, so that S represents $p(\mathbf{x})$ and S' is obtained from S by multiplying by a continuous, non-negative function $h(\mathbf{x})$. Then provided $h(\mathbf{x})p(\mathbf{x})$ is not identically zero, S' represents the probability distribution proportional to $h(\mathbf{x})p(\mathbf{x})$.*

Proof. Let $\mathcal{P}(\mathcal{X})'$ be the space of all finite measures on $\mathcal{X}$, (recall that $\mathcal{P}(\mathcal{X})$ includes only probability distributions) and similarly let $\mathcal{P}(\mathcal{P}(\mathcal{X}))'$ be the space of all finite measures on $\mathcal{P}(\mathcal{X})$. The important abstraction here is that "multiplying by a function h" is an operator on the space $\mathcal{P}(\mathcal{P}(\mathcal{X}))'$. Denote the operator by $h\times$ and observe that its effect on $\Phi \in \mathcal{P}(\mathcal{P}(\mathcal{X}))'$ can be described as follows: for any bounded continuous function $F : \mathcal{P}(\mathcal{X})' \to \mathbb{R}$,

$$(h \times \Phi)F = \int F(h\mu)\Phi(d\mu).$$

(The notation $h\mu$ is the standard multiplication of a real-valued measure by a real-valued function.) If we can show this operator is continuous (for continuous, non-negative h) as a function $\mathcal{P}(\mathcal{P}(\mathcal{X}))' \to \mathcal{P}(\mathcal{P}(\mathcal{X}))'$ then we are done. Indeed, let $S_1, S_2, \ldots$ be the elements of S, which by the definition of a particle set tend to $\mathbf{p}$. Then

$$(h \times S_n)F = \int F(h\mu)S_n(d\mu) \longrightarrow_{n\to\infty} F(hp(\mathbf{x}))$$

(justified because F and h are continuous so $F(h\,\bullet)$ is also continuous), and since this holds for all F we conclude that $h \times S_n \to h\mathbf{p}$, as desired.

In fact the proof is not complete, since the operator $h\times$ does not correctly normalise the distributions. But this is easily patched up: let $\alpha(\mu)$ be a real-valued function such that $\alpha(\mu)h\mu$ is normalised whenever μ is normalised. Provided hp is not identically zero, one can easily show[5] α is continuous in a neighbourhood of p, so now the argument works as before. ■

Remark. It is a trivial extension to prove the theorem for piecewise continuous h. However, h cannot be too pathological. For instance, the favourite measure theory counterexample $h = 1$ on rationals and 0 on irrationals shows the theorem is not true for all h: the particle set S defined by equation 2.11 only ever takes rational values for the $\mathbf{x}_i^{(n)}$ so $h \times S = S$, whereas $h(\mathbf{x})p(\mathbf{x}) = 0$ since h is zero almost everywhere.

Figure 2.5 shows an example of the effect of multiplying a particle set by a function. Note that the multiplication operation can be used to prove equation 2.12 is a representation of the uniform distribution on $[\delta, 1]$ for $\delta > 0$. Let S be the standard Monte Carlo set for the restriction of the triangular distribution to $[\delta, 1]$ — so S represents a distribution with probability density function $\propto x$. Set $h(x) = 1/x$; then h is a continuous function on $[\delta, 1]$. So by the theorem, $h \times S$ represents a distribution with pdf proportional to $1/x \times x$ — that is, $h \times S$ represents the uniform distribution on $[\delta, 1]$. Finally, it is easy to check that $h \times S$ is precisely the particle set given by equation 2.12.

2.3.2 Applying dynamics

Diffusing a particle set according to some stochastic dynamics is another important step in the Condensation algorithm. The stochastic dynamics take the

[5] Let $\beta = 1/\alpha$, so $\beta(\mu) = \int h\,d\mu$. If $\mu_1, \mu_2, \ldots \to \mu$ then $\beta(\mu_i) = \int h\,d\mu_i \to \int h\,d\mu = \beta(\mu)$.

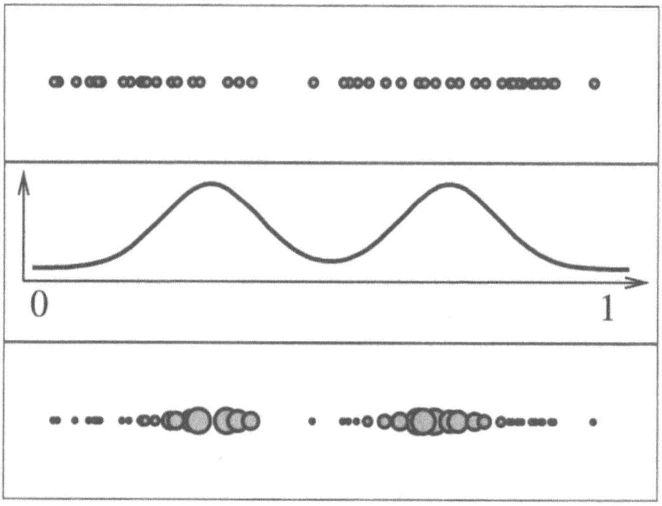

Figure 2.5: ***Particle set multiplied by a function.*** *Top: a realisation of the particle set given by equation 2.5, representing the uniform distribution. Middle: a function $h(\mathbf{x})$ by which the particle set will be multiplied. Bottom: the particle set obtained by multiplying the top particle set by h. Note the particles are in the same positions but their weights have changed.*

form of a conditional density $p(\mathbf{x}'|\mathbf{x})\,d\mathbf{x}'$, which expresses the probability that the target object is now in the volume $d\mathbf{x}'$, given that at the last time-step it was in configuration $\mathbf{x}$.

Definition (Applying dynamics to a particle set) Let S be a particle set on $\mathcal{X}$, with $S = \{(\mathbf{x}_i^{(n)}, \pi_i^{(n)})_{i=1}^n, n = 1, 2, \ldots\}$, and let $p(\mathbf{x}'|\mathbf{x})$ be a conditional density. Define a new particle set, $S' = \{(\mathbf{x}_i^{(n)'}, \pi_i^{(n)'})_{i=1}^n, n = 1, 2, \ldots\}$ by (as usual, the superscript n's are dropped):

$$\begin{aligned} &\mathbf{x}_i' \vdash p(\mathbf{x}_i'|\mathbf{x}_i) \\ &\pi_i' = \pi_i. \end{aligned} \tag{2.13}$$

We say S' is obtained from S by *applying the dynamics* $p(\mathbf{x}'|\mathbf{x})$. If S represents a distribution $p(\mathbf{x})$, this is written on a Condensation diagram as

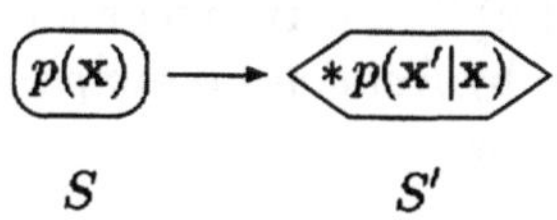

One would expect that applying dynamics diffuses the distribution represented by a particle set, according to the transition kernel $p(\mathbf{x}'|\mathbf{x})$, and this is indeed the case. Unfortunately, the statement of a theorem to this effect is a little more complex, due to a technical difficulty: the above method of applying dynamics is well-defined for particle sets, but not for general members of

$\mathcal{P}(\mathcal{P}(\mathcal{X}))$. The method depends on picking each particle exactly once, and diffusing it, whereas these particles do not necessarily exist for a general member of $\mathcal{P}(\mathcal{P}(\mathcal{X}))$. However, there is a way out: it turns out that if the initial particle set was produced by a resampling operation (defined in the next section) we do get a well-defined operator. Hence this theorem is stated now but proved after the next section.

Theorem 2.3 *Let S, S' be as above, so that S represents $p(\mathbf{x})$ and S' is obtained from S by applying the dynamics $p(\mathbf{x}'|\mathbf{x})$. Suppose in addition that S was produced from some other particle set by a resampling operation. Then S' represents the distribution $p'(\mathbf{x}')$ given by*

$$p'(\mathbf{x}') = \int_{\mathbf{x}} p(\mathbf{x}'|\mathbf{x})p(\mathbf{x})\,d\mathbf{x}.$$

Proof. This is proved as Theorem A.4, in Appendix A.3. ■

Incidentally, we conjecture that this theorem is true without the additional assumption that S is a resampled set.

Figure 2.6 gives an example of applying dynamics to a particle set.

2.3.3 Resampling

Many different particle sets can represent a distribution $p(\mathbf{x})$, but some do so more efficiently than others. Look back to Figure 2.4, where one feels intuitively that the centre panel is the "best" representation of the uniform distribution, while the lower panel is the "worst". This concept of efficiency will be formalised later as effective sample size, but it is mentioned here as the motivation for the resampling operation. It will turn out that a key feature of an efficient representation is that most of the particles have equal weights, and resampling is a way to achieve this.

Definition (Random resampling) Let $S = (\mathbf{x}_i^{(n)}, \pi_i^{(n)})_{i=1}^n, n = 1, 2, \ldots$ be a particle set, and define a new particle set S' by

$$\mathbf{x}_i^{(n)'} = \mathbf{x}_j^{(n)} \text{ with probability } \pi_j^{(n)},$$
$$\pi_i^{(n)'} = 1/n,$$

where the random choice of $\mathbf{x}_i^{(n)'}$ occurs independently for $i = 1, \ldots n$. We say S' has been obtained from S by *(random) resampling*, and depict this operation on a Condensation diagram by

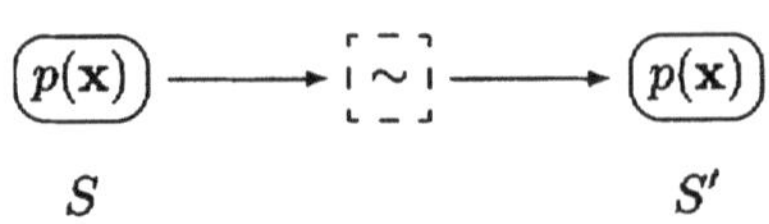

The next theorem explains why $p(\mathbf{x})$ appears twice in this diagram: *resampling does not affect the distribution represented.* In fact, this is the critical

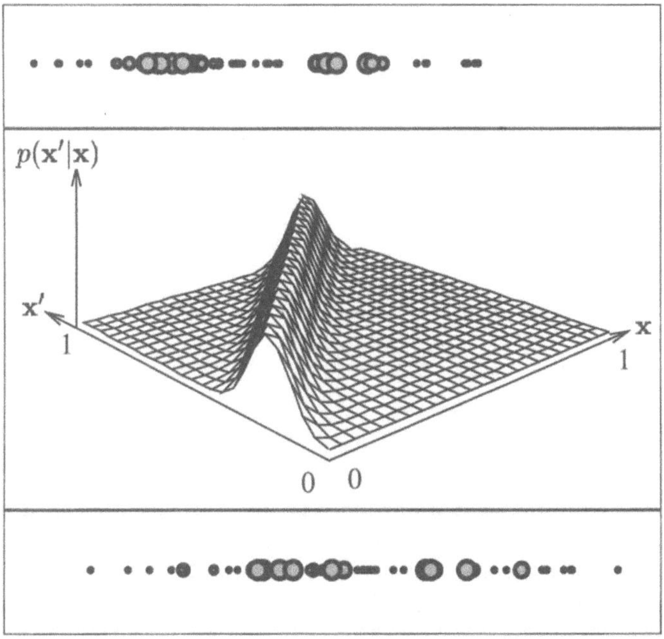

Figure 2.6: ***Applying dynamics to a particle set.*** *Top: a realisation of a particle set representing a bimodal distribution on the unit interval. Middle: the dynamics* $p(\mathbf{x}'|\mathbf{x})$ *to be applied. Note the dynamics incorporate a deterministic shift and some Gaussian diffusion. Bottom: a realisation of the particle set obtained by applying these dynamics. The weights of the particles are unchanged, but their positions have been shifted according to the dynamics: a deterministic shift to the right, plus some Gaussian noise.*

result that will enable us to complete the proof that Condensation is asymptotically correct. It circumvents the independence assumption discussed at the end of Section 2.1, by showing that although the particles in a resampled set are not independent, the set nevertheless represents the same distribution as before.

Theorem 2.4 *Let S' be a particle set obtained from S by random resampling. If S represents a distribution $p(\mathbf{x})$, then so does S'.*

Proof. This is proved as Theorem A.2 of Appendix A.3. There is no real intuition to be gleaned from the method of proof: it consists of some careful abstraction of the resampling concept and an appeal to a technical lemma of Del Moral (1998). ■

Figure 2.7 gives an example. Looking at this figure, one gets the feeling that the middle panel is somehow "more random" than necessary. We know how many times each particle should be selected by the resampling process if

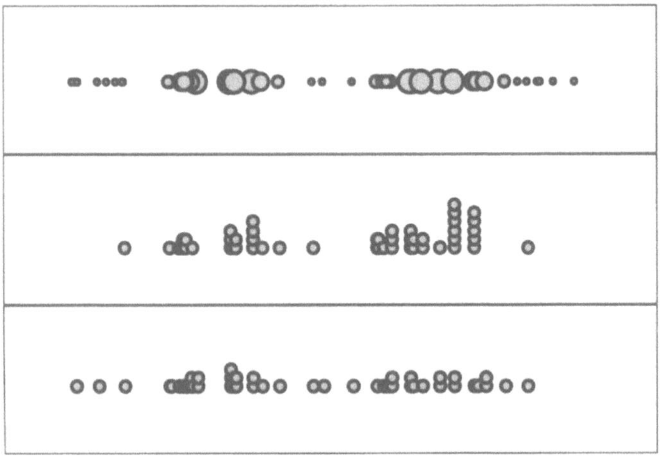

Figure 2.7: ***Resampling a particle set.*** *Top: a realisation of a particle set representing a bimodal distribution on the unit interval. Middle: a realisation of the particle set obtained by random resampling on the original particle set. Some particles are chosen several times by the resampling; they are shown offset in the y-direction. Note that no new particle positions are generated by the resampling, but that some positions "die out" because they are not selected at all from the original set. Bottom: the result of* deterministic *resampling on the original set. Each particle of the original set is replicated by the "right" number of times — proportional to its weight, but rounded to an integer of course. This form of resampling produces a higher effective sample size than random resampling whenever it is valid. However, there are certain pathological cases in which deterministic resampling alters the distribution represented by a particle set.*

it is to resemble the previous set as closely as possible: the ith particle should be chosen $n\pi_i^{(n)}$ times. Because of rounding this cannot always be achieved exactly, but we can certainly guarantee that it holds to the nearest integer number of particles. A procedure that achieves this is called "deterministic resampling", and though many schemes can be used, we adopt the following definition for concreteness.

Definition (Deterministic resampling) Let $S = (\mathbf{x}_i^{(n)}, \pi_i^{(n)})_{i=1}^n, n = 1, 2, \ldots$ be a particle set. Write $c_i^{(n)} = \sum_{j=1}^i \pi_j^{(n)}$ for the cumulative weights, and define a new particle set S' by

$$\mathbf{x}_i^{(n)\prime} = \mathbf{x}_{j^*(i)}^{(n)}, \text{ where } j^*(i) = \text{smallest } j \text{ such that } c_j^{(n)} \geq i/n.$$

$$\pi_i^{(n)\prime} = 1/n.$$

This type of sampling is called deterministic because once the $\mathbf{x}_i$ are known then so are the $\mathbf{x}_i'$. The rule given does indeed ensure that the proportion

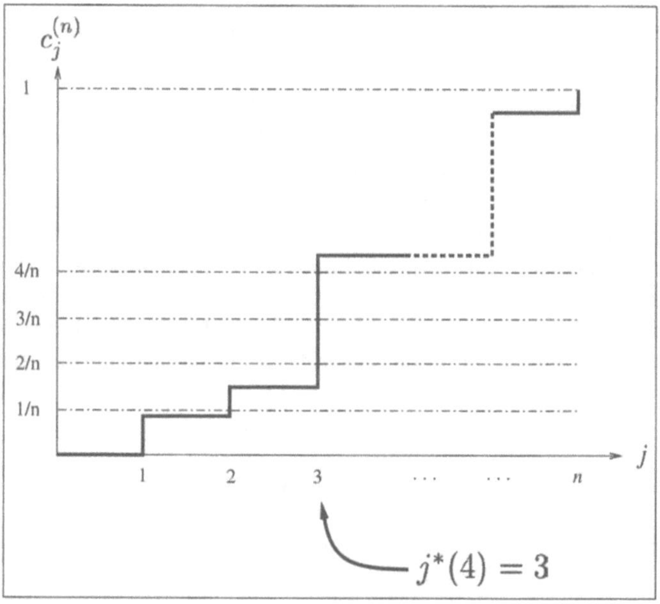

Figure 2.8: ***The rule for deterministic resampling.*** *The graph plots the cumulative weights $c_j^{(n)}$, and shows an example of how to calculate $j^*(i)$, which is defined to be the smallest j such that $c_j^{(n)} \geq i/n$.*

of $\mathbf{x}_i^{(n)\prime}$ equal to a given $\mathbf{x}_j^{(n)}$ is close to $\pi_j^{(n)}$. To see this, think of making n equally-spaced horizontal slices through a graph of the cumulative probabilities $c_i^{(n)}$ (Figure 2.8).

Theorem 2.4 does *not* hold with random sampling replaced by deterministic resampling. For a counterexample, let $0 < \varepsilon < 1$ and define a particle set by

$$\mathbf{x}_i \vdash \begin{cases} \text{Rect}[0, \frac{1}{2}] & \text{when } i \text{ is odd} \\ \text{Rect}[\frac{1}{2}, 1] & \text{when } i \text{ is even} \end{cases}$$

$$\pi_i = \begin{cases} (1-\varepsilon)/n & \text{when } i \text{ is odd} \\ (1+\varepsilon)/n & \text{when } i \text{ is even.} \end{cases}$$

Deterministic resampling of this particle set selects only the $\mathbf{x}_i$ with i even (except possibly for $\mathbf{x}_n$, if n is odd), meaning that all (except possibly for one) resampled particles are in the interval $[\frac{1}{2}, 1]$. Yet the original particle set assigns weight $(1-\varepsilon)/2$ to the interval $[0, \frac{1}{2}]$.

Fortunately, this counterexample is pathological. We have never encountered a real problem in which deterministic resampling alters the distribution represented by a particle set. In the remainder of this book, we will assume the particle sets arising in tracking problems can be resampled deterministically without altering the distribution represented. Resampling symbols in

Condensation diagrams will mean deterministic resampling unless it is stated otherwise.

Remark. This discussion goes some way to addressing a question of Bill Triggs[6] about Condensation. His question can be rephrased as "Condensation is essentially performing numerical integration on complex probability distributions. Why then do you need *random* numbers to perform Condensation?" The answer is that in principle one doesn't *need* random numbers; it's a matter of designing the most appropriate combination of deterministic and stochastic operations on particle sets. For example, it would seem that applying dynamics is always easier by a genuinely stochastic method, whereas for resampling operations deterministic methods are preferable except in certain pathological cases like the one above.

2.4 The Condensation theorem

Our present objective is to prove that the Condensation diagram of equation 2.4 produces a particle set that represents the right-hand side of equation 2.2. In fact, we would like more than this: we also want to be able to iterate equation 2.4 a finite number of times and still be confident of its asymptotic behaviour. This will justify the use of the Condensation algorithm for tracking, since we know that by choosing n sufficiently large, a simulation of the nth element of the particle sets will approximate arbitrarily well the T Bayesian calculations like equation 2.2 that we originally wanted to perform.

The next theorem says this is indeed possible. Call a Condensation diagram *well-formed* if every application of dynamics is immediately preceded by a random resampling. Then we have

Theorem 2.5 (The Condensation theorem) ***Informal statement:*** *any well-formed Condensation diagram produces a particle set representing the distribution you would expect.*

Formal statement: *let $O_1, O_2, \ldots O_T$ be a finite sequence of operations on particle sets consisting of either random resampling, multiplying by some non-negative continuous function, or application of continuous dynamics. Let $P_1, \ldots P_T$ be the operators on probability distributions that correspond to these operations (resample $\equiv$ identity; multiply by h $\equiv$ 'multiply by h and renormalise'; apply dynamics $p(\mathbf{x}'|\mathbf{x}) \equiv$ 'weight by $p(\mathbf{x}'|\mathbf{x})$ and integrate w.r.t. $\mathbf{x}$'). Moreover suppose that whenever O_n is the application of some dynamics then O_{n-1} is a resampling. Let S be an initial particle set representing a distribution $p(\mathbf{x})$. Then $S' = O_T \circ \ldots \circ O_1 S$ is a particle set representing $p' = P_T \circ \ldots \circ P_1 p$.*

Proof. This is a trivial induction using Theorems 2.2, 2.3 and 2.4. The necessity for a Condensation diagram to be well-formed is because of the assumption made in Theorem 2.3. ∎

[6]See the discussion following Blake et al. (1998)

Del Moral's (1998) result, as interpreted and applied in this chapter, amounts to the first correct proof of the asymptotic validity of the Condensation algorithm in the computer vision literature. As mentioned at the end of Section 2.1, the proof in Isard and Blake (1998a) uses an induction based on the formulation of factored sampling given by Grenander (see the discussion in the next section). Unfortunately, this formulation requires the particles to be independent samples from the prior for each time-step, whereas it is clear that after resampling the particles are not independent.

2.5 The relation to factored sampling, or "where did the proof go?"

The basic objective of Monte Carlo methods[7] is to evaluate the integral of a function $f(\mathbf{x})$ over a space $\mathcal{X}$ with respect to a measure (or prior, if you are thinking statistically) $p(\mathbf{x})$:

$$I = \int_{\mathbf{x}\in\mathcal{X}} f(\mathbf{x})dp(\mathbf{x}).$$

The simplest Monte Carlo method estimates I by drawing n i.i.d. samples $\mathbf{x}_1, \ldots \mathbf{x}_n$ from the distribution p then setting

$$\widehat{I} = \frac{1}{n}\sum_{i=1}^{n} f(\mathbf{x}_i).$$

In the language of particle sets, this is just the estimator based on the standard Monte Carlo set for p, multiplied by the function f. In the specific case that p is a prior and f a likelihood function which can be evaluated up to constant, this method is termed *factored sampling*, an idiom which appears to have been introduced by Ulf Grenander in his early work on pattern theory (Grenander, 1981).[8] A precise statement of the factored sampling method from the point of view of particle sets is given in the next theorem.

Theorem 2.6 (Factored sampling) *Let $p(\mathbf{x})$ be a probability density (the prior) on $\mathcal{X}$, and $f(\mathbf{x})$ a real-valued function (the likelihood) on $\mathcal{X}$. Define the posterior p' by $p'(\mathbf{x}) = f(\mathbf{x})p(\mathbf{x})/Z$, where Z is an unknown constant chosen so that p' is a probability density on $\mathcal{X}$. Then if S is any particle set representing p, $f \times S$ is a particle set representing p'.*

Proof. This is a restatement of Theorem 2.2, page 17. Just replace f with h and ignore the normalisation constant. ∎

[7] Mackay (1999) is a remarkably lucid, brief introduction to the basic ideas of Monte Carlo methods.

[8] Mysteriously, he has been known to disown responsibility for this terminology in recent years.

Previous proofs of this result (e.g. Grenander, 1981, p. 596) used probabilistic methods involving an appeal to the law of large numbers, whereas Theorem 2.2 relies only on elementary analysis, admittedly in the rather abstract space $\mathcal{P}(\mathcal{P}(\mathcal{X}))$. Moreover, theorem 2.6 is actually more general than the formulation in Grenander (1976–1981) since S can be *any* particle set which represents p, whereas Grenander requires the samples from p to be independent. It is natural to wonder, therefore, "where did the proof go?" The answer is that it has been subsumed in our definition of what it means for a particle set to represent a probability distribution. We defined our way out of the difficult corner, which is showing that a discrete sample of a continuous distribution approximates the continuous distribution as the size of the sample grows. When factored sampling is actually implemented, some kind of sampling is involved, and to prove its asymptotic correctness requires Proposition 2.1, or more generally, the Condensation theorem (Theorem 2.5). These results, in turn, depend on Lemma A.1 of Appendix A.3, which, as Del Moral (1998) points out, "can be regarded as a uniform weak law of large numbers".

2.6 "Good" particle sets and the effective sample size

The ultimate goal of a particle set is to estimate some useful quantity: for concreteness, say we want to know the x coordinate of the centroid of a tracked object. Let $S = (\mathbf{x}_i^{(n)}, \pi_i^{(n)})$ be a particle set representing a distribution $p(\mathbf{x})$ over a configuration space $\mathcal{X}$, and let $\phi : \mathcal{X} \to \mathbb{R}$ be a function whose expectation is the "useful quantity" we are interested in. That is, we want to estimate

$$E\phi = \int_{\mathcal{X}} \phi(\mathbf{x})p(\mathbf{x})\,d\mathbf{x}.$$

This quantity can be estimated from S in the obvious way:

Definition (nth estimator of ϕ using S) The nth estimator of ϕ using S is a random variable

$$S_\phi^{(n)} = \sum_{i=1}^{n} \phi(\mathbf{x}_i^{(n)})\pi_i^{(n)}. \tag{2.14}$$

Note that $S_\phi^{(n)}$ really is a random variable, since the $\mathbf{x}_i^{(n)}$ and $\pi_i^{(n)}$ are. Thus it has a mean $ES_\phi^{(n)}$ and a variance $\text{var}(S_\phi^{(n)})$.

For continuous functions ϕ, it can be shown[9] that

$$\lim_{n\to\infty} ES_\phi^{(n)} = E\phi. \tag{2.15}$$

Assume for a moment that the convergence here is reasonably rapid. Then we are still not assured that $S_\phi^{(n)}$ is a good estimator of $E\phi$, since $\text{var}(S_\phi^{(n)})$ might

[9] Let $\varphi : \mathcal{P}(\mathcal{X}) \to \mathbb{R}; p \mapsto E_p\phi$. Note φ is continuous since ϕ is. Since $S_n \to \mathbf{p}$, we have $\lim_{n\to\infty} S_n\varphi = \mathbf{p}\varphi = E_\mathbf{p}\phi$, where the first step follows because φ is continuous. But $S_n\varphi = ES_n$, so we are done.

be large. To assess this, we compare $\mathrm{var}(S_\phi^{(n)})$ with $\mathrm{var}(\phi)$ using the *effective sample size*:

Definition (Effective sample size) Let $\mathcal{E}_\phi$ be any estimator of $E\phi$. The *effective sample size* of $\mathcal{E}_\phi$, denoted $\mathcal{N}(\mathcal{E}_\phi)$, is given by

$$\mathcal{N}(\mathcal{E}_\phi) = \frac{\mathrm{var}(\phi)}{\mathrm{var}(\mathcal{E}_\phi)}, \tag{2.16}$$

where $\mathrm{var}(\phi) = \int_{\mathbf{x}} (\phi(\mathbf{x}) - E\phi)^2 p(\mathbf{x})\, d\mathbf{x}$.

This definition follows Carpenter et al. (1997) and Doucet (1998); Geweke (1989) calls this quantity the relative numerical efficiency. The reason for this terminology will be clear after the following definition.

Definition (nth Monte Carlo estimator of ϕ) Let $Y_1, Y_2, \ldots$ be i.i.d. random variables with distribution $p(\mathbf{x})$. The nth Monte Carlo estimator of ϕ is

$$M_\phi^{(n)} = \frac{1}{n} \sum_{i=1}^{n} \phi(Y_i).$$

An equivalent definition would be to let S be the standard Monte Carlo set for p, define $M = \phi \times S$, and take $M_\phi^{(n)}$ to be the nth estimator of ϕ using M as defined above.

By elementary probability theory, $EM_\phi^{(n)} = E\phi$, and $\mathrm{var}(M_\phi^{(n)}) = \mathrm{var}(\phi)/n$. Thus the effective sample size of the nth Monte Carlo estimator of ϕ is just n:

$$\mathcal{N}(M_\phi^{(n)}) = \frac{\mathrm{var}(\phi)}{\mathrm{var}(\phi)/n} = n.$$

This explains the choice of terminology: if $\mathcal{N}(\mathcal{E}_\phi) = n$ for some estimator $\mathcal{E}_\phi$, then the same variance could have been achieved by using a Monte Carlo estimator with n samples.

For most estimators $\mathcal{E}_\phi$, there is no way to calculate $\mathcal{N}(\mathcal{E}_\phi)$ exactly. The quantity $\mathrm{var}(\mathcal{E}_\phi)$ must almost always be estimated, and often $\mathrm{var}(\phi)$ must be estimated too. To remind us of exactly how many things are being estimated, we adopt the notations

$$\widehat{\mathcal{N}} = \frac{\mathrm{var}(\phi)}{\text{estimate for } \mathrm{var}(\mathcal{E}_\phi)}$$
$$\widehat{\widehat{\mathcal{N}}} = \frac{\text{estimate for } \mathrm{var}(\phi)}{\text{estimate for } \mathrm{var}(\mathcal{E}_\phi)}.$$

In the specific case of the particle set S, the estimate for $\mathrm{var}(S_\phi^{(n)})$ is obtained as the variance of m draws from $S_\phi^{(n)}$ — call them $s_\phi^{(n)}(1), \ldots s_\phi^{(n)}(m)$. Each draw from $S_\phi^{(n)}$ involves an independent realisation of S; by dropping superscript ns we can write the realisations of the m simulations of S as

$$(\mathbf{x}_{i,1}, \pi_{i,1})_{i=1}^{n}, (\mathbf{x}_{i,2}, \pi_{i,2})_{i=1}^{n}, \ldots (\mathbf{x}_{i,m}, \pi_{i,m})_{i=1}^{n},$$

and express the $s_\phi^{(n)}$ as

$$s_\phi^{(n)}(k) = \sum_{i=1}^{n} \phi(\mathbf{x}_{i,k})\pi_{i,k}.$$

Now follow your nose, making the obvious definitions

$$\widehat{E\phi^2} = \frac{1}{m}\sum_{i=1}^{n}\sum_{k=1}^{m} \phi(\mathbf{x}_{i,k})^2\pi_{i,k}$$

$$\widehat{E\phi} = \frac{1}{m}\sum_{i=1}^{n}\sum_{k=1}^{m} \phi(\mathbf{x}_{i,k})\pi_{i,k}$$

$$\text{estimate of } \operatorname{var}(\phi) = \frac{m}{m-1}(\widehat{E\phi^2} - (\widehat{E\phi})^2)$$

$$\text{estimate of } \operatorname{var}(S_\phi^{(n)}) = \frac{1}{m-1}\sum_{k=1}^{m}(s_\phi^{(n)}(k) - \widehat{E\phi})^2$$

and finally the effective sample size based on two estimates:

$$\widehat{\widehat{\mathcal{N}}}(S_\phi^{(n)}) = \frac{m\left((\widehat{E\phi^2} - (\widehat{E\phi})^2)\right)}{\sum_{k=1}^{m}(s_\phi^{(n)}(k) - E\phi)^2}. \tag{2.17}$$

Of course if $E\phi$ and $\operatorname{var}(\phi)$ are known, then we use the effective sample size based on one estimate:

$$\widehat{\mathcal{N}}(S_\phi^{(n)}) = \frac{m \operatorname{var}(\phi)}{\sum_{k=1}^{m}(s_\phi^{(n)}(k) - E\phi)^2}. \tag{2.18}$$

Example Effective sample sizes were estimated for the three particle sets of Figure 2.4 with $\phi(\mathbf{x}) = \mathbf{x}$; the results are graphed in Figure 2.9. An important point to note is that because the particle set defined by equation 2.11 is completely deterministic, the variance of its estimator is zero and its effective sample size is infinite. To show a more meaningful graph, we replaced half of the particles in this set with random uniform variates; as the figure shows, the particle set still has $\widehat{\mathcal{N}}(S_\phi^{(n)}) > n$. The estimates $\widehat{\mathcal{N}}$ are based on $m = 100$ runs of each particle set (for each value of n); if $\widehat{\mathcal{N}}$ was a perfect estimator of $\mathcal{N}$ then every line on the graph would be straight. Hence we see that even in this simple example and with $m = 100$ runs, the estimates $\widehat{\mathcal{N}}$ are rather noisy; the problems will only get worse in real tracking problems where the computational burden of 100 runs is too high. This is a good reason to seek for a less costly method of assessing the quality of our particle sets. The next section introduces just such a method: the *survival diagnostic*.

2.6.1 The survival diagnostic

As remarked above, it is never actually possible to calculate the effective sample size in real problems — it must be estimated by repeated runs of a stochastic

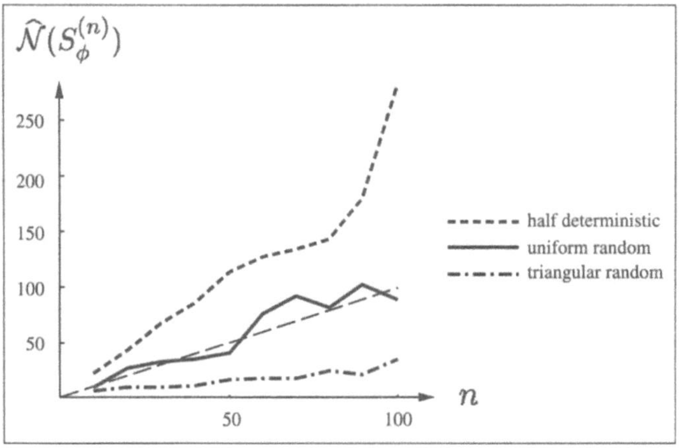

Figure 2.9: ***Estimated effective sample sizes for three particle sets.*** *$\widehat{\mathcal{N}}(S_\phi^{(n)})$ is plotted for each of the particle sets in Figure 2.4, except that the deterministic particle set (corresponding to the middle panel of Figure 2.4, and labelled "half deterministic" here) is altered so that every second particle is drawn randomly from the uniform distribution — otherwise the effective sample size for this particle set would be infinite. The function being estimated is $\phi(\mathbf{x}) = \mathbf{x}$, so the particle sets are simply estimating the mean of a uniform distribution on the unit interval. The "uniform random" set, corresponding to the top panel of Figure 2.4, is in fact the standard Monte Carlo estimator, so by definition its effective sample size is $\mathcal{N} = n$. For comparison, this is shown as a light dashed line; note that $\widehat{\mathcal{N}}$ is a fairly noisy estimator of $\mathcal{N}$ but that it appears to be unbiased. The "half deterministic" particle set has $\widehat{\mathcal{N}} > n$, reflecting the fact that half of the particles in this set perform exact numerical integration of $\phi(\mathbf{x})p(\mathbf{x})$. The "triangular random" particle set (corresponding to the lowest panel of Figure 2.4) has $\widehat{\mathcal{N}} < n$, as we would expect for a stochastic particle set with unequal weights. Each value of $\widehat{\mathcal{N}}$ here was calculated using $m = 100$ runs of the particle set.*

algorithm, or by some other statistical technique. One useful estimator is the "survival diagnostic". The survival diagnostic has some limitations but is useful because it is so easy to calculate.

Definition (Survival diagnostic) Let $S = (\mathbf{x}_i^{(n)}, \pi_i^{(n)})$ be a particle set. The *survival diagnostic* $\mathcal{D}(n)$ is a random variable defined by

$$\mathcal{D}(n) = \left(\sum_{i=1}^{n} \pi_i^{(n)2} \right)^{-1} . \qquad (2.19)$$

The intuitive interpretation of the survival diagnostic is as the number of "useful" particles in a particle set. A useful particle is one that makes a significant contribution when a quantity ϕ is estimated as in equation 2.14. If, say, k of the particles have weights of almost $1/k$ and the remaining particles have weights an order of magnitude or more smaller, then it's clear that only the k heavy particles make a meaningful contribution to equation 2.14. This can be directly related to effective sample size: if the k heavy particles are essentially independent draws from the distribution represented, then we would expect the effective sample size of such a particle set to be about k. In general, of course, the k heavy particles will not be independent and the effective sample size would be even less. What is the connection to the survival diagnostic? Let us formalise the special case just discussed, so that k of n particles have weight $1/k$ and the remaining $n-k$ particles have weight zero. Then direct calculation gives

$$\mathcal{D}(n) = \left(\sum_{i=1}^{k}(1/k)^2\right)^{-1} = k,$$

which agrees nicely with the fact that k is a tight upper bound on the effective sample size above, with equality holding if the particles are independent. This also reveals why $\mathcal{D}$ is called a "survival" diagnostic: $\mathcal{D}$ gives an indication as to how many particles will survive a resampling operation.

It is intuitively clear that, in the general case (when particles can have weights other than 0 or $1/k$), $\mathcal{D}(n)$ is still a useful indicator of the effective sample size. In fact Liu and Chen (1995, 1998), Liu yciteLiu96 and Kong et al. (1994) have proved this although their approximations are often not valid for the types of particle set encountered in tracking problems. Nevertheless, it is interesting to see the formal connection between $\mathcal{D}(n)$ and $\mathcal{N}(n)$, and this is explained in the next section. First, however, let us consider one example.

Example The "triangular" particle set (2.12) representing the uniform distribution on the unit interval was used to test the survival diagnostic. Figure 2.10 shows the results for various values of n. The estimated effective sample size $\widehat{\mathcal{N}}(n)$ was calculated and plotted just as in Figure 2.9, from $m = 100$ simulations of the particle set. This time, however, it was done not just for $\phi(x) = x$ but also for $\phi(x) = x^2$. For each of the 100 simulations, we obtain a value of $\mathcal{D}(n)$ via equation 2.19; the median, maximum, and minimum of these 100 values are also plotted for various values of n. The key points to note are:

- On average, the survival diagnostic overestimates the effective sample size (but appears to do so by a roughly constant multiple).
- The value of the survival diagnostic on any particular realisation of the particle set ranges from near zero to about $0.9n$; in particular, it will occasionally have a value less than the effective sample size.

Nevertheless, the survival diagnostic is a useful tool for analysing the performance of different Condensation techniques. It can give warning of imminent failure for instance: a low value of $\mathcal{D}$ (say, below 10–20 in a real tracking problem) means almost certain trouble, since (regardless of the actual value of $\mathcal{N}$, which might be greater than $\mathcal{D}$ as Figure 2.10 shows) this means few particles will survive the next resampling. Moreover, the computational effort of

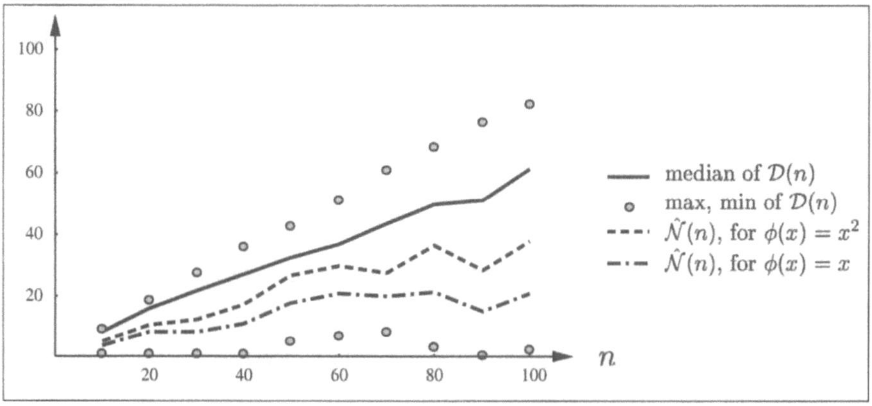

Figure 2.10: ***Survival diagnostic compared to effective sample size.*** *The survival diagnostic $\mathcal{D}(n)$ and estimated effective sample size $\widehat{\mathcal{N}}(n)$ (for two different choices of ϕ) are plotted for the "triangular" particle set (equation 2.12). The estimates are based on $m = 100$ simulations; hence the plots for $\mathcal{D}(n)$ summarise 100 different values. Note that, on average, $\mathcal{D}$ overestimates $\mathcal{N}$ but that it does so rather noisily.*

evaluating $\mathcal{D}$ is negligible, so it is certainly realistic to monitor the survival diagnostic constantly.

2.6.2 From effective sample size to survival diagnostic

This section explains why $\mathcal{D}(n)$ can be regarded as an estimator of $\mathcal{N}(n)$, under certain assumptions. Suppose that $p(\mathbf{x}), q(\mathbf{x})$ are two probability distributions on the configuration space $\mathcal{X}$. Let S be the standard Monte Carlo particle set for p:

$$\begin{aligned} \mathbf{x}_i^{(n)} &\vdash p(\mathbf{x}), \\ \pi_i^{(n)} &= 1/n. \end{aligned} \tag{2.20}$$

By Proposition 2.1, S represents p. Define another particle set S' by

$$\begin{aligned} \mathbf{x}_i^{(n)'} &\vdash q(\mathbf{x}), \\ \pi_i^{(n)'} &\propto p(\mathbf{x}_i^{(n)'})/q(\mathbf{x}_i^{(n)'}). \end{aligned} \tag{2.21}$$

From one point of view, this is the standard factored sampling scenario considered by Grenander (Section 2.5) with prior q and likelihood p/q, so that S' represents $q \times p/q = p$. However, this S' is more simply viewed as the standard Monte Carlo set for q, multiplied (in the particle set sense) by the function p/q; again we conclude that S' represents the distribution p. Note that in much of the Monte Carlo literature, q would be called the "proposal density" and p the posterior.

As usual, let ϕ be the quantity to be estimated via equation 2.14, so that

$$S_\phi^{(n)} = \frac{1}{n}\sum_{i=1}^{n}\phi(\mathbf{x}_i^{(n)}),$$
$$S'^{(n)}_\phi = \sum_{i=1}^{n}\phi(\mathbf{x}_i^{(n)'})\pi_i^{(n)'}.$$

A result of Kong et al. (1994) states that *provided ϕ varies slowly with* $\mathbf{x}$

$$\frac{\mathrm{var}_q(S'^{(n)}_\phi)}{\mathrm{var}_p(S_\phi^{(n)})} \approx 1 + \mathrm{var}_q(p(\mathbf{x})/q(\mathbf{x})). \tag{2.22}$$

The right-hand side here can be re-expressed:

$$\begin{aligned}
1+\mathrm{var}_q(p(\mathbf{x})/q(\mathbf{x})) &= E_q\left(\frac{p(\mathbf{x})}{q(\mathbf{x})}\right)^2 && \text{by a simple calculation}\\
&\approx \tfrac{1}{n}\textstyle\sum_{i=1}^{n}\left(\frac{p\left(\mathbf{x}_i^{(n)'}\right)}{q\left(\mathbf{x}_i^{(n)'}\right)}\right)^2 && \text{as the } \mathbf{x}_i^{(n)'} \text{ are drawn from } q\\
&\approx n\textstyle\sum_{i=1}^{n}\pi_i^{(n)'2} && \text{since } p(\mathbf{x}_i^{(n)'})/q(\mathbf{x}_i^{(n)'}) \approx n\pi_i^{(n)'}\\
& && \text{(Cor. 2.8 below)}\\
&= n/\mathcal{D}(n) && \text{by the definition of } \mathcal{D}.
\end{aligned} \tag{2.23}$$

Finally, observe that by elementary probability theory,

$$\mathrm{var}_p(S_\phi^{(n)}) = \frac{1}{n}\mathrm{var}_p(\phi), \tag{2.24}$$

so that

$$\begin{aligned}
\mathcal{N}(S'^{(n)}_\phi) &= \frac{\mathrm{var}_p(\phi)}{\mathrm{var}_q(S'^{(n)}_\phi)} && \text{definition of } \mathcal{N}\text{, (2.16)}\\
&= \frac{n\mathrm{var}_p(S_\phi^{(n)})}{\mathrm{var}_q(S'^{(n)}_\phi)} && \text{by (2.24)}\\
&\approx n/(1+\mathrm{var}_q(p(\mathbf{x})/q(\mathbf{x}))) && \text{by (2.22), provided } \phi \text{ slowly varying}\\
&\approx \mathcal{D}(n) && \text{by (2.23).}
\end{aligned}$$

That establishes the connection between the effective sample size $\mathcal{N}$ and the survival diagnostic $\mathcal{D}$: if all our approximations are valid, then $\mathcal{D}$ is a good estimator of $\mathcal{N}$. However, there are two reasons that in practice this estimate cannot be relied upon:

- The condition for equation 2.22 to hold, that $\phi(\mathbf{x})$ must vary slowly with $\mathbf{x}$, can be violated arbitrarily badly even by everyday practical examples. Carpenter et al. (1999) give a lucid demonstration of this in their appendix.

- The derivation was for a particle set S' consisting of *independent* samples from the proposal density. However, the Condensation algorithm uses the resampling operation which introduces dependencies between the particles. That is why $\mathcal{D}$ is a meaningless estimate of $\mathcal{N}$ immediately after resampling: all the weights are equal, so $\mathcal{D}(n) = n$, whereas the true value of $\mathcal{N}(n)$ can be no greater than before the resampling.

This explains the difference between the graphs of $\mathcal{D}$ and $\mathcal{N}$ in Figure 2.10. As already remarked, however, the survival diagnostic remains a very useful tool for monitoring tracker performance.

2.6.3 Estimating the weight normalisation

Here we reprove a well-known result from the Monte Carlo literature (e.g. Geweke, 1989; Stewart, 1983) in the framework of particle sets. This concerns the normalisation constant for the weights in certain standard particle sets; the result was used in the previous section and it will also be useful later.

Lemma 2.7 *Let S be the standard Monte Carlo particle set (2.10) representing a distribution $p(\mathbf{x})$, and let $q(\mathbf{x})$ be another probability distribution with the same support as p.*

Then $\frac{1}{n}\sum_{i=1}^{n} q(\mathbf{x}_i^{(n)})/p(\mathbf{x}_i^{(n)})$ is a random variable tending weakly to the constant value 1 as $n \to \infty$.

Proof. Let X be a random variable with law $p(\mathbf{x})$ and set $Y = q(X)/p(X)$. Let $Y_1, Y_2, \ldots$ be i.i.d. with same law as Y. Then $1/n\sum^n Y_i \to EY$ by the law of large numbers, and we have $EY = \int (q(\mathbf{x})/p(\mathbf{x})\, p(\mathbf{x})\, d\mathbf{x} = 1$. ■

The Lemma can be restated in the factored sampling context:

Corollary 2.8 *Let S' be the standard "factored sampling" particle set (2.21) representing p with proposal density q, and let $Z_n = \sum_{i=1}^{n} p(\mathbf{x}_i^{(n)'})/q(\mathbf{x}_i^{(n)'})$ be the normalising constant for the weights. Then Z_n/n tends weakly to 1 as $n \to \infty$. Informally, we have*

$$\pi_i^{(n)'} \approx \frac{1}{n}\frac{p(\mathbf{x}_i^{(n)'})}{q(\mathbf{x}_i^{(n)'})}.$$

Proof. This is just a rephrasing of Lemma 2.7 (it looks like the ps and qs are round the wrong way, but they aren't): consider yet another particle set S'' which is the standard Monte Carlo set representing q; now apply the lemma. ■

2.6.4 Effective sample size of a resampled set

Since the resampling operation never generates particles in new configurations, it is clear that the effective sample size of a particle set after resampling is never greater than before the resampling. However, it is difficult to obtain more explicit results than this, except in certain special cases. One such case is when the particle set being resampled already has equal weights. (Of course, you would never actually *want* to resample a set which already has equal weights, since the whole point of resampling is to make the weights more equal! But one can think of this as a sort of worst case.) The basic result is summarised by the following

Slogan: randomly resampling an equally-weighted set of independent particles halves the effective sample size.

To see why this is true, let $X_1, \ldots X_n$ be i.i.d. random variables with law $p(\mathbf{x})$, and Y be the estimator of their mean given by

$$Y = \frac{1}{n}\sum_{i=1}^{n} X_i.$$

Without loss of generality, $EX = 0$, so we have

$$\begin{aligned} \mathrm{var}(X) &= EX^2 := \sigma^2 \\ EY &= EX = 0 \\ \mathrm{var}(Y) &= \frac{\sigma^2}{n}. \end{aligned}$$

We can think of the X_i as an equally-weighted particle set consisting of independent particles — the standard Monte Carlo set for p. Then the result of resampling the set is obtained by setting

$$X_i^* = X_j \text{ with probability } 1/n$$

and the estimator of the mean of the resampled set is given by

$$Y^* = \frac{1}{n}\sum_{i=1}^{n} X_i^*.$$

Clearly $EY^* = EX^* = 0$. To find the effective sample size of the resampled set, we first need to calculate $\mathrm{var}(Y^*)$. We have

$$\mathrm{var}(Y^*) = \mathrm{var}(E(Y^*|X_1, \ldots X_n)) + E(\mathrm{var}(Y^*|X_1, \ldots X_n)). \tag{2.25}$$

The first term here is just var(Y), or σ^2/n, while the second term can be calculated as follows:

$$
\begin{aligned}
E(\text{var}(Y^*|X_1,\ldots X_n)) &= E\left(\text{var}(1/n\sum X_i^*|X_1,\ldots X_n)\right)\\
&= \frac{1}{n^2}E\left(\text{var}(\sum X_i^*|X_1,\ldots X_n)\right)\\
&= \frac{1}{n}E\left(\text{var}(X_1^*|X_1,\ldots X_n)\right)\\
&= \frac{1}{n}E\left(1/n\sum(X_i-Y)^2\right)\\
&= \frac{1}{n^2}E\left(\sum(X_i^2-2X_iY+Y^2)\right)\\
&= \frac{1}{n^2}\sum(EX_i^2-2E(X_iY)+EY^2)\\
&= \frac{1}{n^2}\sum(\sigma^2-2\sigma^2/n+\sigma^2/n)\\
&= \frac{1}{n^2}\sum(\sigma^2-\sigma^2/n)\\
&= \frac{\sigma^2}{n}(1-1/n).
\end{aligned}
$$

Hence

$$
\frac{\text{var}(Y^*)}{\text{var}(Y)} = \frac{\sigma^2/n+\sigma^2/n(1-1/n)}{\sigma^2/n} = 1+1-1/n = 2-1/n. \qquad (2.26)
$$

Thus, for large n, the resampled estimator has twice the variance of the original estimator — in other words, resampling has halved the effective sample size of the particle set.

Note that this result is a particular case of the rule of thumb given by Liu and Chen (1995). This states that if $\tilde{\mu}$ is the resampled estimator and $\hat{\mu}$ the original estimator, and W is a random variable representing the weights of the original particle set, then

$$
\frac{\text{var}(\tilde{\mu})}{\text{var}(\hat{\mu})} \approx \frac{n-1}{n} + \frac{1}{1+\text{var}(W)}. \qquad (2.27)
$$

Since in this case the weights were assumed to be equal, $W \equiv 1$ so that var(W) $= 0$ and equation 2.27 reduces to equation 2.26.

2.7 A brief history of Condensation

Particle filters have been used in a diverse range of applied sciences, and the basic algorithms were discovered independently by researchers in several of these disciplines. The generality and ease of implementation of particle filters make them ideal for many simulation and signal processing problems. In physics, particle filters have been used for fluid mechanics (e.g. McKean, 1967) and statistical mechanics (e.g. Liggett, 1985), where they go by the name of *interacting particle systems*. Statisticians have used particle filters for various types of data analysis (e.g. Gordon et al., 1993; Rubin, 1988; Smith and Gelfand, 1992; Carpenter et al., 1999) and there have also been applications in econometrics (e.g. Pitt and Shepherd, 1997). Signal processing is another fruitful field for particle filters; Doucet (1998) gives many references to applications in this arena. Important contributions to particle filtering from a measure-theoretic point of view have included Del Moral (1998), whose result on resampling was used to underpin this chapter, Crisan and Lyons (1997, 1999), and Crisan et al. (1998). Finally, the field of computer vision — with its intractable likelihood functions and often complex configuration spaces — has seen an explosion of interest in particle filters since Isard and Blake first introduced the Condensation algorithm in 1996. Some important contributions have been those of Isard and Blake (1998a), Black and Jepson (1998), and Heap and Hogg (1998). This book presents two more advances: a probabilistic exclusion principle (Chapter 6) which builds on the discrete-continuous formulation of Condensation (Isard and Blake, 1998c) and partitioned sampling (Chapter 7) which makes Condensation feasible for multiple targets and extended configuration spaces.

The plethora of fields using particle filters and the frequent reinvention of the main ideas make an accurate history of particle filtering a difficult if not impossible task. Doucet (1998) and Isard (1998) both make brave attempts, so we confine ourselves here to some brief comments. Handschin and Mayne (1969, 1970) were one set of researchers to originate a method of sequential Bayesian filtering based on random sampling. However, their formulation did not involve *re*sampling, so the effective sample size of their particle sets would have decreased rapidly to zero for any of the applications that interest us. An algorithm incorporating resampling was popularised by Rubin (1988) under the name "Sampling Importance Resampling" (SIR) and was put to good use by Smith and Gelfand (1992). Gordon et al. (1993) and Kitagawa (1996) independently applied SIR to time series data filtering. Gordon's algorithm is called the "bootstrap filter", while Kitagawa (who uses the generic term "Monte Carlo filter") appears to have been the first non-physicist to describe the random samples as "particles".

In computer vision, Isard and Blake (1996) rediscovered the principles behind SIR, and by applying these to the active contours framework, which includes learned models of deformation and shape spaces, exhibited excellent results for tracking objects in considerable clutter. By adding a discrete variable to the state space (Isard and Blake, 1998c), they later extended the tracking to include motion classification. Heap and Hogg (1998) used a similar framework to cope with shape changes which are not easily modelled by the shape spaces of Blake and Isard (1998), including changes in the topology of the contour. Black and Jepson (1998) tackled the problem of gesture recognition using Condensation on optical flow measurements. Black and Fleet (1999) were able to detect motion discontinuities accurately by another method which also employed Condensation.

2.8 Some alternatives to Condensation

The most easily-implemented and well-known alternative to Condensation is the Kalman filter (Bozic, 1979). As Isard and Blake (1998a) emphasise in their original Condensation paper, the Kalman filter makes Gaussian assumptions on the likelihood functions which are completely inappropriate in contour tracking problems in which the background contains clutter; hence the Kalman filter fails irretrievably in these situations. Some interesting recent work which demonstrates the same problem for singularities and end effects is Deutscher et al. (1999).

Another way of stating the reason for the failure of Kalman filters in clutter is that the Kalman filter can maintain only a unimodal probability distribution for the current state, whereas when tracking in clutter it is essential to consider multiple hypotheses. The extended Kalman filter (Gelb, 1974) does little to rectify this problem, since it relies on a linearisation of a non-linear measurement model which is still not flexible enough to entertain multiple hypotheses.

Perhaps the most frequently-asked question one is asked about Condensation is "Wouldn't it be easier to use a mixture of Gaussians?" It is certainly true that a mixture-of-Gaussians approach permits multi-modal densities to be propagated over time, but an exact implementation requires an exponentially increasing number of Gaussians in the mixture. Of course, many of these Gaussians have low weights, so it is possible to prune the mixture using various heuristics without affecting the posterior too much. However, the real problem is how to propagate the centres of the Gaussians. *If* a tractable form of the likelihood function is available — in particular, if there is a way to rapidly find the peaks of the likelihood and the width of these peaks — there are sensible ways of doing this. But the fact is that the likelihood functions in contour tracking problems (see Chapter 3) are not at all tractable; they have narrow peaks at unknown locations which must be detected in spaces of up to, say, 40 dimensions. The stochastic dynamics employed in a particle filter are the best method I am aware of for locating the peaks in these likelihood functions.

For true devotees of Gaussian mixtures, another way of looking at this is to regard Condensation as a crude form of mixture-of-Gaussians. By placing a Gaussian kernel with a default covariance centred at each particle in a particle set, one obtains a Gaussian mixture model. The resampling operation then becomes the pruning mechanism, but because of the Condensation theorem, we are guaranteed that our pruning "heuristic" is asymptotically correct.

3

Contour likelihoods

Image processing operations often produce a list of detected features. To perform inference on these outputs, we need a probabilistic description of how these features are generated. Such descriptions are often called *generative models*. Printed character recognition provides a simple example (Hinton et al., 1992): we can think of an actual typeset character on the page as the result of combining a perfect copy of the character with some ink spatter. A generative model for printed characters must therefore specify the ideal version of the character and the details of how ink spatter is produced. The type of generative model used in image analysis depends on the application. Mumford and Gidas (1999) give a useful taxonomy into models for image textures (Geman et al., 1990; Zhu et al., 1997), for decomposition of an image into regions (Geman and Geman, 1984; Mumford and Shah, 1989), for grammatical parsing of shapes (Fu, 1982), for template matching, for disparity maps, for specific tasks such as face recognition (Hallinan et al., 1999; Leung et al., 1995), and for the application that interests us here, namely contours (Grenander et al., 1991a; Mumford, 1993). The choice of model must be governed both by tractability (can we compute with it, and can we do so sufficiently fast?) and by realism (does the model accurately capture the essential features for our application?). Since one of our goals is to perform inference on video sequences in real time, ease of computation is crucial.

3.1 A generative model for image features

Recall our paradigm for measuring images, introduced in Section 1.2.3. Only a set of certain line segments in the image is analysed; the chosen line segments are called measurement lines, and some simple image processing is performed on each measurement line to determine the positions of some features which correspond to edges in the image. For each feature there are precisely three possibilities: (1) the feature corresponds to the edge of a target object (these features are useful to us — they can be used to infer the configuration of a target); (2) the feature corresponds to something in the *interior* of the target (these features may be useful to us, depending on how much we know about what the interior of the target should look like); (3) the feature corresponds to part of the background (these features are not useful to us — they are distractions). Our generative model must specify how all three types of features arise, and involves the following fundamental assumptions.

Edge of target object Suppose a given measurement line of length L intersects the target object's boundary, as in Figure 3.1(a). Then we expect a feature to be reported near this boundary. The feature will not, however, be *precisely* at the intersection of the target boundary and the measurement line. The main reason for this is that our model of the target's shape is idealised: the target position $\mathbf{x}$ is an element of a shape space which only approximates the actual observed shape of the object in an image.[1] If z is the position of the detected feature, and ν the position of the intersection (Figure 3.1(a)), our generative model makes the following assumption.

- **Assumption 1** z is drawn randomly from a density $\mathcal{G}(z|\nu)$, called the *boundary density function.*

Throughout the book we take $\mathcal{G}$ to be a truncated Gaussian centred on ν:

$$\mathcal{G}(z|\nu) = \begin{cases} \text{const} \times \exp(-\frac{1}{2}\frac{(z-\nu)^2}{\sigma^2}) & \text{if } z \in [0, L], \\ 0 & \text{otherwise,} \end{cases}$$

where the constant is chosen so that $\mathcal{G}$ integrates to 1, and σ is estimated from a template by the maximum likelihood technique described in North (1998). A typical value is $\sigma = 7$ pixels.

Edge of target object: non-detection probability Suppose again that a given measurement line intersects the target object's boundary. Occasionally, as in Figure 3.1(b), it will happen that no feature corresponding to this intersection is detected. This could be caused by the background having insufficient contrast with the target. Our simplest generative model assumes that

- **Assumption 2** There is a fixed probability q_{01} that a single target boundary feature will not be detected.

The notation q_{01} is a mnemonic for the probability of detecting zero target boundary features out of the one which was expected. Thus $q_{11} = 1 - q_{01}$ is the probability of detecting one target boundary feature when one is expected. Chapter 6 will consider the possibility of more than one target boundary occurring on a measurement line, and Chapter 5 explains

[1] Examples of this type of approximation include the use of 2D affine transformations for a distant rigid planar object moving in 3D, or more generally assuming only rigid transformations for an object which may exhibit small non-rigid deformations.

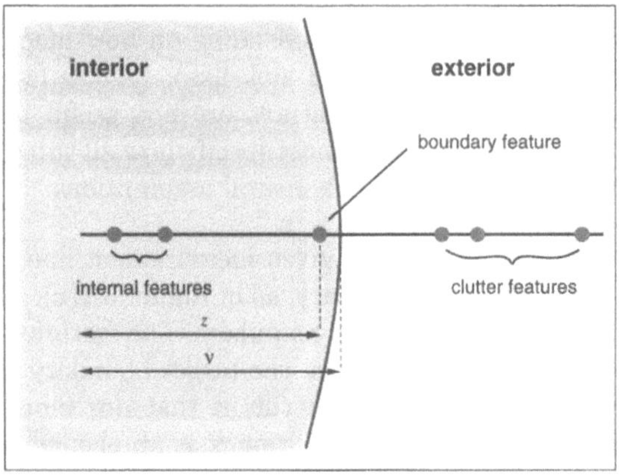

(a) Target boundary present and detected

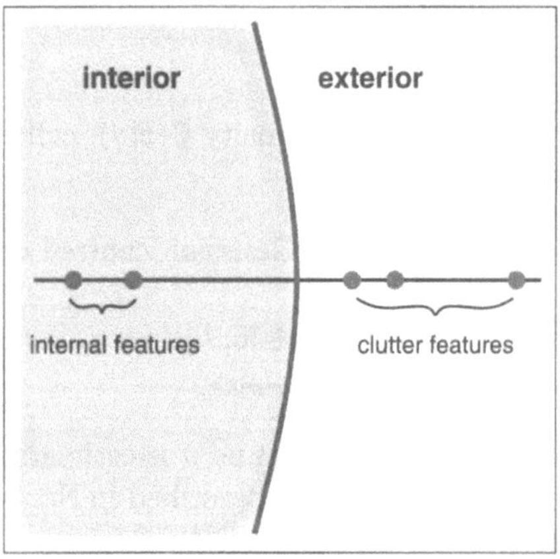

(b) Target boundary present but not detected

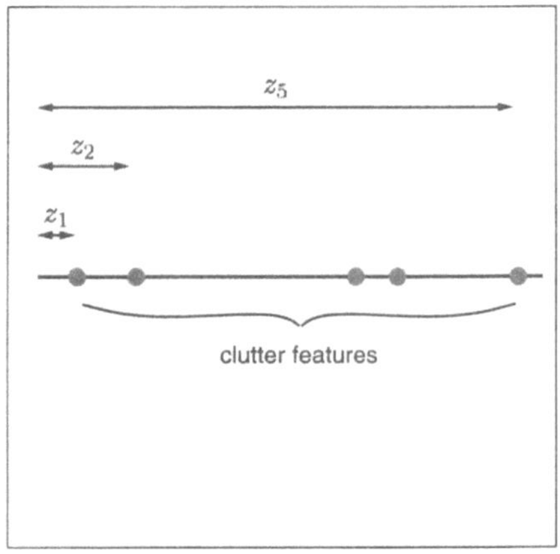

(c) Target not present: "clutter"

Figure 3.1: ***Fundamental assumptions for the generative model.*** *Each diagram is a schematic close-up of just one measurement line, demonstrating one of the three fundamental assumptions in the generative model. The solid circles are features detected on the measurement line — in practice these would be, for example, the locations of maximal grey-scale gradients over a given threshold. (a) The vertical curve is a part of the target object's* idealised *boundary, with the interior of the object being on the left. In general we expect to find a feature near the intersection of this boundary with the measurement line; the precise position of such a feature is generated by drawing from a distribution* $\mathcal{G}(z|\nu)$. *(b) However, occasionally the boundary of the target will not be detected as a feature (perhaps because the grey-scale of the background is too similar to the target at that point). This event is modelled by a* non-detection probability q_{01}. *(c) If no target is present, all features arise from random background clutter. Our simplest generative model states there is a probability* $b(n)$ *that* n *features are detected, and the positions of these features are uniformly distributed on the measurement line.*

how we can allow "non-detection" events to incorporate occlusions, by allowing for statistical dependence between the measurement lines.

Distribution of background clutter features If a measurement line does not intersect a target, and lies entirely in the background (Figure 3.1(c)), then any features detected on it are assumed to be generated by a generic background distribution. Suitable background models are discussed in Section 3.2.1, but for now we specify the two basic assumptions for the simplest generative model:

- **Assumption 3** There is a fixed probability $b(n)$ of detecting n clutter features on a measurement line. In general, $b(n)$ depends on the length L of a measurement line, and to emphasise this we sometimes write $b_L(n)$.
- **Assumption 4** The positions $z_1, z_2, \ldots z_n$ are uniformly distributed over the measurement line.

Distribution of interior target features Often, part or all of a measurement line lies in the *interior* of a target (Figure 3.1(a) and (b)). Clearly, it would be a good idea to have a separate generative model for features detected on the interior portions of measurement lines, but this must be based on knowledge of the target's appearance. Sometimes this is not available, and we instead adopt the following assumption.

- **Assumption 5** Interior features are generated by the same model as background features.

An intuitive justification for this would be: "The generative model for the background describes the occurrence of features on all objects in the universe. The target is an object in the universe, so the same generative model should describe occurrences of features on it." Nevertheless, the target has a special status in our formulation of the universe, so it could be argued that it deserves a special model; Section 3.1.3 addresses this by describing one such model for the target interior.

Independence of distinct measurement lines Features detected on the 1D measurement lines are a function of a 2D spatial process, the image itself. Thus we need to specify the statistical relationship between features on different measurement lines. The simplest generative model makes the following assumption.

- **Assumption 6** Feature outputs on distinct measurement lines are statistically independent.

Section 3.2.2 explains why this assumption is appropriate for the problems tackled in this book. Essentially, it can be shown that features on measurement lines separated by a normal displacement of more than 30 pixels have negligible correlation. Chapter 5 gives the details of a more sophisticated model which permits dependence between the measurement lines.

3.1.1 The generic contour likelihood

In this section we formalise the notions of the generative model introduced in the previous section, and derive a probability density function for features produced by the generative model. First, the concepts of "intersection number" and "intersection innovation" are introduced.

Definition (Intersection number) Let $\mathbf{x} \in \mathcal{X}$ be an element of the shape space, and let $\mathcal{L}_1, \ldots \mathcal{L}_M$ be the M measurement lines to be processed for a given image. Then the *intersection number* of $\mathcal{L}_i$ with $\mathbf{x}$, written $c_{\mathbf{x}}(i)$, is just the number of points at which the line segment $\mathcal{L}_i$ intersects the contour corresponding to $\mathbf{x}$. If $\mathbf{x}$ or i is clear from the context, it can be omitted.

Definition (Intersection innovation) Suppose the measurement line $\mathcal{L}_i$ has intersection number $c_{\mathbf{x}}(i)$ with configuration $\mathbf{x}$. Then the *intersection innovation* of $\mathcal{L}_i$ with $\mathbf{x}$, written $\boldsymbol{\nu}_{\mathbf{x}}(i)$, is just the set of Euclidean distances from the intersections to the start of the measurement line.[2] Again, if either $\mathbf{x}$ or i is clear from the context, it can be omitted. Note that if $c_{\mathbf{x}}(i) = 0$, then $\boldsymbol{\nu}_{\mathbf{x}}(i) = \emptyset$, the empty set. If $c_{\mathbf{x}}(i) = n > 0$, we often write $\boldsymbol{\nu}$ as $\{\nu_1, \ldots \nu_n\}$.

Example Figure 3.2 gives an example showing intersection numbers and innovations.

The pdf for the generative model will be derived first for the output $\mathbf{z}$ of just *one* measurement line of length L. For a given target configuration $\mathbf{x}$, there are two possibilities we wish to consider: the measurement line may have intersection number $c = 0$ or $c = 1$. (In principle, higher intersection numbers are possible but this is very rare for short measurement lines used in practice and we prefer to neglect the possibility at this stage. Chapter 6 considers $c > 1$ in the more natural setting of multiple targets.) For the simplest case, in which the detection probabilities $q_{cc} = 1$, these densities are denoted $p_c(n; \mathbf{z})$, or $p_c(n; \mathbf{z}|\boldsymbol{\nu})$ when the intersection innovation is non-empty. The following proposition gives formulas for the p_c.

Proposition 3.1 *Suppose the non-detection probability $q_{01} = 0$. Then the pdfs for the generative model on a single measurement line of length L are given by*

$$p_0(n; \mathbf{z}) = b(n)/L^n$$

$$p_1(n; \mathbf{z}|\boldsymbol{\nu} = \{\nu_1\}) = b(n-1)\sum_{k=1}^{n} \mathcal{G}(z_k|\nu_1)/nL^{n-1}. \tag{3.1}$$

Remarks. Because of the discrete parameter n which indicates how many arguments z_i follow, the functions p_c are not quite pdfs in the standard sense. However, this is a technical detail which can be avoided by explaining the notation more clearly. For example, $p_0(n; z_1, \ldots z_n|\boldsymbol{\nu})$ is just shorthand for $p_0(z_1, \ldots z_n|n, \boldsymbol{\nu})p(n)$, so that $p_0(n; z_1, \ldots z_n)dz_1 \ldots dz_n$ is just the probability

[2] Recall from the footnote on page 7 that measurement lines have an orientation, so the "start" of the line has a well-defined meaning here.

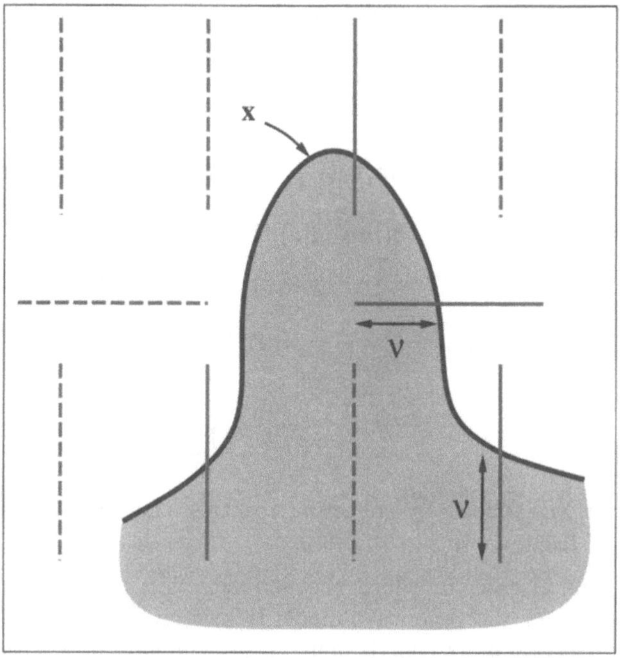

Figure 3.2: ***Intersection numbers.*** *A sparse set of measurement lines is shown near a contour* $\mathbf{x}$*. Dotted lines have intersection number* $c_{\mathbf{x}} = 0$ *with* $\mathbf{x}$*, whereas solid lines have intersection number* $c_{\mathbf{x}} = 1$*. Intersection innovations* ν *are shown for two of the measurement lines. (In practice, the set of measurement lines would be more dense than shown here — see Section 3.2.3.)*

of obtaining n features *and* that these features lie in the volume $dz_1 \ldots dz_n$ centred on $\mathbf{z} = (z_1, \ldots z_n)$.

Another subtle point is that each z_i is a point in the image, which would normally be described by an x and y coordinate. However, in this context the features are constrained to lie on the measurement line, which is a one-dimensional subset of the image. So the notation dz_i refers to a small one-dimensional subset of the image.

Proof. The formula for p_0 follows almost immediately from the assumptions. By definition there is a probability $b(n)$ of obtaining n features, and these are distributed uniformly on the length L of the measurement line. Hence $p_0(n; z_1, \ldots z_n) = b(n)/L^n$.

The formula for p_1 relies on a simple combinatoric argument. First note the generative model described above is equivalent to the following: (i) The number of background features, say m, is selected with probability $b(m)$ (Assumption 3). (ii) The positions of the background features are drawn from the uniform distribution on the measurement line, obtaining say $b_1, \ldots b_m$ (Assumption 4). (iii) The position a of the boundary feature is selected by a random draw from $\mathcal{G}(a|\nu)$ (Assumption 1). (iv) The total number of features n is set to $m+1$, and the vector $(a, b_1, \dots b_m)$ is randomly permuted and reported as $(z_1, \dots z_n)$. In mathematical terms, we can say that a permutation ρ is selected uniformly at random from the symmetric group $\mathbb{S}_n$, and applied to the vector $(a, b_1, \dots b_m)$.

After stage (iii), the pdf $p(m; a, b_1, \dots b_m|\nu)$ of the unpermuted vector is just

$$b(m)\mathcal{G}(a|\nu)/L^m,$$

and since each of the $n!$ permutations has an equal probability we calculate

$$\begin{aligned} p_1(n; z_1, \dots z_n|\boldsymbol{\nu} = \{\nu_1\}) &= b(m) \sum_{\rho \in \mathbb{S}_n} \frac{\mathcal{G}(z_{\rho(1)}|\boldsymbol{\nu})}{L^m} \times \frac{1}{n!} \\ &= b(n-1) \sum_{k=1}^{n} \frac{\mathcal{G}(z_k|\nu_1)}{nL^{n-1}}, \end{aligned}$$

where the last line follows by collecting together the $(n-1)!$ permutations which leave z_k fixed. ∎

For the general case, in which the non-detection probability q_{01} can be positive, the pdf for the generative model is written as $\tilde{p}_c(n; \mathbf{z})$, or $\tilde{p}_c(n; \mathbf{z}|\boldsymbol{\nu})$ when the intersection innovation is non-empty. The next proposition gives the appropriate formulas.

Proposition 3.2 *The pdfs for the generative model on a single measurement line of length L are given by*

$$\begin{aligned} \tilde{p}_0(n; \mathbf{z}) &= p_0(n; \mathbf{z}) \\ \tilde{p}_1(n; \mathbf{z}|\boldsymbol{\nu}) &= q_{01}p_0(n; \mathbf{z}) + q_{11}p_1(n; \mathbf{z}|\boldsymbol{\nu}). \end{aligned} \tag{3.2}$$

Proof. Invoking Assumption 2 about the "non-detection event", this is a trivial application of conditional probability to the pdfs in Proposition 3.1. ∎

A typical graph of $\tilde{p}_1(n; \mathbf{z}|\boldsymbol{\nu} = \{\nu\})$ considered as a likelihood function (i.e. fixed $n, \mathbf{z}$ and varying ν) is shown in Figure 3.3.

Of course, Propositions 3.1 and 3.2 discuss the case of only one measurement line, whereas we need a density for the measurements $\mathbf{Z} = (\mathbf{z}^{(1)}, \dots \mathbf{z}^{(M)})$ on M measurement lines. However, our assumption of independence between measurement lines makes this an elementary extension. This is confirmed by the next proposition, which also serves as a definition of the *generic likelihood function.*

Proposition 3.3 (Generic likelihood function) *Suppose a set of measurement lines $\mathcal{L}_1, \dots \mathcal{L}_M$ has intersection numbers $c_{\mathbf{x}}(1), \dots c_{\mathbf{x}}(M)$ and intersection innovations $\boldsymbol{\nu}_{\mathbf{x}}(1), \dots \boldsymbol{\nu}_{\mathbf{x}}(M)$ with a configuration $\mathbf{x} \in \mathcal{X}$. Then the* generic likelihood function *for the measurements $\mathbf{Z} = (\mathbf{z}^{(1)}, \dots \mathbf{z}^{(M)})$ is*

$$\mathcal{P}_{\text{gen}}(\mathbf{Z}|\mathbf{x}) = \prod_{i=1}^{M} \tilde{p}_{c_{\mathbf{x}}(i)}(\mathbf{z}^{(i)}|\boldsymbol{\nu}_{\mathbf{x}}(i)). \tag{3.3}$$

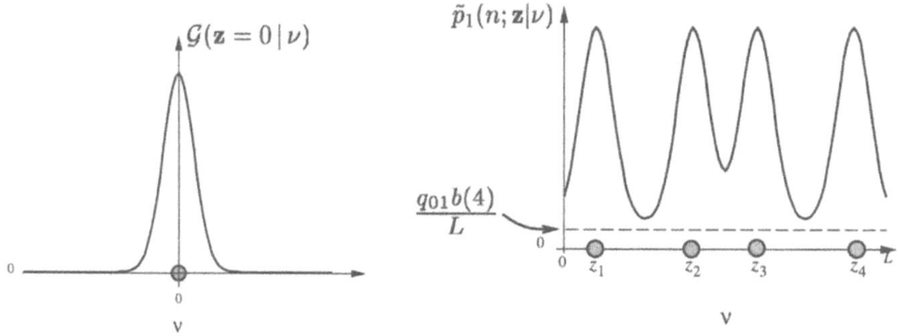

Figure 3.3: ***The one-target likelihood function for a single measurement line.*** *Left: the boundary feature distribution $\mathcal{G}(z = 0|\nu)$. Right: the one-target likelihood function $\tilde{p}_1(n;\mathbf{z}|\nu)$ graphed with respect to ν. The likelihood is a linear combination of shifted copies of $\mathcal{G}(z|\cdot)$ and of the constant p_0. It peaks near the four measurements z_i (shown as shaded circles).*

Proof. This follows immediately from Assumption 6 (independence of measurement lines) and the pdfs for single measurement lines given in Proposition 3.2. ∎

3.1.2 The Poisson likelihood

Section 3.2.1 will discuss background models in more detail, but it will be useful to introduce one special case now: the spatial Poisson process. If the background features (and, by Assumption 5, the interior features) arise from a Poisson process then the likelihood of the measurements takes on a particularly simple form. To see this, adopt the following assumption.

- **Assumption 7** The background features on a one-dimensional measurement line obey a Poisson law with density λ. That is,

$$b_L(n) = e^{-\lambda L}\frac{(\lambda L)^n}{n!}.$$

Then it is easy to check that the densities for a single measurement line (equation 3.2) become

$$\tilde{p}_0(n;\mathbf{z}) = \frac{e^{-\lambda L}\lambda^n}{n!},$$

$$\tilde{p}_1(n;\mathbf{z}|\boldsymbol{\nu} = \{\nu_1\}) = \frac{e^{-\lambda L}\lambda^n}{n!}\left(q_{01} + \frac{q_{11}}{\lambda}\sum_{k=1}^{n}\frac{1}{\sqrt{2\pi}\sigma}\exp(-\frac{1}{2}\frac{(z_k-\nu_1)^2}{\sigma^2})\right),$$

where the boundary density function $\mathcal{G}$ has been explicitly written out as Gaussian.[3] Thus with the extra Assumption 7, the generic likelihood function $\mathcal{P}_{\text{gen}}$

[3] A minor approximation was adopted here: recall that the boundary density function is actually a *truncated* Gaussian, so the normalisation constant is not quite equal to $1/\sqrt{2\pi}\sigma$.

defined in equation 3.3 becomes the *Poisson likelihood function*:

$$\mathcal{P}_{\text{poisson}}(\mathbf{Z}|\mathbf{x}) = \left(\prod_{i=1}^{M} \frac{e^{-\lambda L} \lambda^{n_i}}{n_i!} \right) \times \prod_{j \text{ s.t. } c_{\mathbf{x}}(j)=1} \left(q_{01} + \frac{q_{11}}{\lambda} \sum_{k=1}^{n_i} \frac{1}{\sqrt{2\pi}\sigma} \exp(-\frac{1}{2} \frac{(z_k^{(j)} - \nu_{\mathbf{x},1}(j))^2}{\sigma^2}) \right). \tag{3.4}$$

(Here n_i is the number of features detected on the ith measurement line. As remarked on page 42, we continue to neglect the possibility that $c_{\mathbf{x}}(j) > 1$.) The importance of this expression is that it can be computed almost effortlessly. The first factor is a constant (remember, we are treating this function as a likelihood, so $\mathbf{Z}$ is fixed and $\mathbf{x}$ varies) and need not be computed at all. The second factor is easy to compute as it is, but one can gain even more speed by approximating the sum by its largest element — with $\sigma \approx 7$ pixels it is very rare for more than one feature to contribute significantly. Writing $d_{\text{min}}(i)$ for the distance on the ith measurement line from the target intersection to the nearest feature, this leaves us with

$$\mathcal{P}_{\text{poisson}}(\mathbf{Z}|\mathbf{x}) = \text{const.} \times \prod_{\substack{\text{meas. lines } i \\ \text{intersecting } \mathbf{x}}} \left(q_{01} + \frac{q_{11}}{\sqrt{2\pi}\sigma\lambda} \exp(-\frac{1}{2} \frac{d_{\text{min}}(i)^2}{\sigma^2}) \right).$$

The numbers appearing in the product can be cached for a suitable discretisation of d_{min}, so that the log-likelihood can be evaluated (up to a constant) very rapidly, as the sum of at most M pre-computed constants.

Note that the Poisson likelihood $\mathcal{P}_{\text{poisson}}$ is precisely the one used by Isard and Blake (1998a) in the original Condensation paper.

3.1.3 The interior-exterior likelihood

One of the assumptions adopted to derive the generic likelihood $\mathcal{P}_{\text{gen}}$ was that features detected in the *interior* of an opaque target obey the same generative model as those in the exterior (Assumption 5, page 41). In the absence of a more specific model of the target interior, this is reasonable enough, but one feels that useful information is being thrown away by adopting such a generic approach. So this section attempts to answer the question: is there a simple way to model the target interior? The answer is yes. Moreover, by adopting the model suggested here, the power of the likelihood function to discriminate targets from clutter is increased.

The proposed model for the interior replaces Assumption 5 by the following.

- **Assumption 8** Interior features are generated by a Poisson process. The density of the process varies slowly enough that it can be treated as constant over the length of a measurement line.

Of course the true value of this constant can be calculated numerically if desired.

For a concrete example, look back to Figure 3.1(a) and (b). Our new assumption means that the features labelled there as "internal" are generated by a Poisson process whose density depends on the location of the measurement line relative to the target. Note that we do allow a different density on each measurement line — the Poisson density is only assumed constant along the interior portion of each single measurement line. This is important in the cases considered later. For example, if the template is the outline of a head, as in Figure 4.4 (page 74), the subject's eyes cause features to be detected on certain measurement lines and this is incorporated into the interior model.

Focus now on a single measurement line of length L, intersecting the target $\mathbf{x}$ at a point where the density of the internal feature process has been learnt as μ. (A simple way to learn μ is described later.) Suppose a length a of the measurement line is *actually* in the interior of the target. Write $f_a(m)$ for the probability that m features are detected on the interior portion of the measurement line. The mnemonic here is that f stands for foreground; compare with $b(n)$ (introduced in Assumption 3), which stands for background. By our new assumption, we have

$$f_a(m) = \exp(-a\mu)\frac{(a\mu)^m}{m!}.$$

Note that the generative model for a measurement line with intersection number $c = 1$ and intersection innovation $\boldsymbol{\nu} = \{\nu\}$ is now equivalent to

1. Draw a randomly from $\mathcal{G}(a|\nu)$. (So a is the reported innovation of the target.)

2. Draw d randomly from the set $\{\texttt{True}, \texttt{False}\}$ with probabilities q_{11}, q_{01} respectively. (So d expresses whether or not the boundary was detected).

3. Draw m randomly from $f_a(m)$, and draw $\phi_1, \ldots \phi_m$ from Rect$[0, a]$. (So m is the number of interior features, and $\phi_1, \ldots \phi_m$ are their innovations.)

4. Draw n randomly from $b_{L-a}(n)$, and draw $\beta_1, \ldots \beta_n$ from Rect$[a, L]$. (So n is the number of exterior (clutter) features, and $\beta_1, \ldots \beta_n$ are their innovations.)

5. If d is True,

   (a) Set $N = n + m + 1$. (So N is the total number of features found.)

   (b) Reorder $(a, \phi_1, \ldots \phi_m, \beta_1, \ldots \beta_n)$ as $(z_1, \ldots z_N)$ by a random permutation.

   otherwise

   (a) Set $N = n + m$. (So N is the total number of features found.)

   (b) Reorder $(\phi_1, \ldots \phi_m, \beta_1, \ldots \beta_n)$ as $(z_1, \ldots z_N)$ by a random permutation.

6. Report $(N; z_1, z_2, \ldots z_N)$.

This is called the *interior-exterior generative model.* The density function corresponding to this model is called the interior-exterior likelihood, and an explicit formula for it is given by the next proposition.

Proposition 3.4 (Interior-exterior likelihood) *Suppose that a given measurement line has length L, intersection number $c = 1$, intersection innovation ν and features generated by the interior-exterior model above. Let $p_1^{\text{ie}}(N;\mathbf{z}|\nu)$ be the likelihood for these features. For a fixed $\mathbf{z} = (z_1, \ldots z_N)$, choose indices i_m such that $z_{i_1} \le z_{i_2} \le \ldots \le z_{i_N}$. Without loss of generality, the measurement line is oriented so that it begins inside the target and ends outside it (so that, for example, z_{i_1} is the most interior feature). Then*

(a) The likelihood given the boundary was detected is

$$p_1^{\text{ie}}(N;\mathbf{z}|\nu, d = \texttt{True}) = \sum_{m=0}^{N-1} \frac{f_{z_{i_{m+1}}}(m)\, b_{L-z_{i_{m+1}}}(N-m-1)\, \mathcal{G}(z_{i_{m+1}}|\nu)}{\binom{N}{m}(N-m)(L-z_{i_{m+1}})^{N-m-1}(z_{i_{m+1}})^m}.$$

(b) The likelihood given the boundary was not detected is

$$p_1^{\text{ie}}(N;\mathbf{z}|\nu, d = \texttt{False}) = \sum_{m=0}^{N} \frac{1}{\binom{N}{m}} \int_{z_{i_m}}^{z_{i_{m+1}}} \frac{f_a(m)\, b_{L-a}(N-m)\, \mathcal{G}(a|\nu)}{(L-a)^{N-m}(a)^m}\, da,$$

where $z_{i_0} := 0$ and $z_{i_{N+1}} := L$.

(c) The unconditioned likelihood is

$$p_1^{\text{ie}}(N;\mathbf{z}|\nu) = q_{01}\, p_1^{\text{ie}}(N;\mathbf{z}|\nu, d = \texttt{False}) + q_{11}\, p_1^{\text{ie}}(N;\mathbf{z}|\nu, d = \texttt{True}).$$

If instead the measurement line has intersection number $c = 0$, then the likelihood p_0^{ie} is the same as the generic likelihood $\tilde{p}_0$ (equation 3.2).

Proof. (a) and (b) are proved in Appendix A.2, and (c) is a trivial consequence. For the final part, just observe that when the crossing number is zero, the generative models for equation 3.2 and p_0^{ie} are the same. ■

By invoking the assumption of independence between measurement lines (Assumption 6), the likelihood for multiple measurement lines can be defined in the same way as equation 3.3. Specifically, define the *interior-exterior likelihood function* for measurements $\mathbf{Z} = (\mathbf{z}^{(1)}, \ldots \mathbf{z}^{(m)})$ as

$$\mathcal{P}_{\text{int-ext}}(\mathbf{Z}|\mathbf{x}) = \prod_{i=1}^{M} p_{c_{\mathbf{x}}(i)}^{\text{ie}}(\mathbf{z}^{(i)}|\nu_{\mathbf{x}}(i)). \tag{3.5}$$

3.1.4 The order statistic likelihood

This section introduces yet another likelihood for contours: the *order statistic likelihood.* The motivation for this is threefold:

1. To obtain a likelihood based on the same assumptions as the generic likelihood (Assumptions 1–6), but with a simpler functional form.

2. To demonstrate a method of deriving likelihoods which does not depend on the distribution of background features being uniform (Assumption 4).

3. To obtain a likelihood for which the dimensionality of the measurements is fixed.

This is achieved by *ignoring* some of the measurement information. Consider our feature detector, which for each measurement line reports the number n of features found and their positions $\mathbf{z} = (z_1, \ldots z_n)$. Instead of this, imagine that for a measurement line of length L, the detector reports *only*:

- the number of features found, n, and

- the displacement $z = d_{\text{min}}$ of the most central feature from the centre of the line:

$$d_{\text{min}} = \arg\min_i |z_i - L/2|.$$

That explains the name of this section: the unique z reported by the feature detector is actually d_{min}, the *first order statistic* of all detected features (when they are ordered by the absolute value of their displacement from the centre of the measurement line). The likelihood functions for the reported values $(n; z)$ are denoted as in Section 3.1.1: $p_c(n; z|\boldsymbol{\nu})$ is the likelihood for a measurement line with intersection number c and intersection innovations $\boldsymbol{\nu}$. As in the rest of this chapter, we defer the case $c > 1$ to Chapter 6; this means that $\boldsymbol{\nu}$ is either the empty set (if $c = 0$), or a singleton $\{\nu\}$ (if $c = 1$). The next proposition gives formulas for the order statistic likelihood function. It is convenient to shift the standard variables for this proposition, writing $z' = z - L/2$ and $\nu' = \nu - L/2$.

Proposition 3.5 (Order statistic likelihood function) *Suppose a feature detector reports the number of features n and first order statistic z described above on a measurement line of length L. Also suppose the features obey the generic generative model of Section 3.1.1, except that $q_{01} = 0$, so that the target boundary feature is always detected. Then if the intersection number of the line is zero, the likelihood is given by*

$$p_0(n; z') = \begin{cases} b(0) & \textit{if } n = 0 \\ \frac{2nb(n)}{L}(1 - 2z'/L)^{n-1} & \textit{if } n \geq 1. \end{cases}$$

If the intersection number is one, then for the particular case that $\nu' = 0$ we have

$$p_1(n; z'|\nu' = 0) = \begin{cases} 0 & \text{if } n = 0 \\ b(0)\mathcal{G}(z'|\nu' = 0) & \text{if } n = 1 \\ b(n)\left(1 - \frac{2z'}{L}\right)^{n-2}\left(\left(1 - \frac{2z'}{L}\right)\mathcal{G}(z'|\nu' = 0) + \right. & \\ \qquad \left. \frac{2(n-1)}{L}\left(1 - \int_{-z'}^{z'} \mathcal{G}(z|\nu = 0)\,dz\right)\right) & \text{if } n \geq 2. \end{cases}$$

Proof. The standard way to calculate the distribution of order statistics is to find their cumulative density functions (cdfs) and then differentiate (Ripley, 1987).

In this case it is a rather routine calculation. When $c = 0$, all n features are distributed uniformly on $[0, L]$ so $\text{Prob}(z' \geq y) = (1 - 2y/L)^n$. Thus the cdf $C(y) = \text{Prob}(z' \leq y) = 1 - (1 - 2y/L)^n$, and differentiating gives the result. The case $c = 1, n = 0$ has zero probability since a target boundary is present and we have assumed a non-detection probability of zero. When $c = 1$ and $n = 1$, z' is distributed as $\mathcal{G}$ by assumption so we are done. That leaves the case $c = 1, n \geq 2$. In this case, $\text{Prob}(z' \geq y) = (1 - 2y/L)^{n-1}\left(1 - \int_{-y}^{y} \mathcal{G}(y')dy'\right)$. The cdf $C(y) = \text{Prob}(z' \leq y) = 1 - \text{Prob}(z' \geq y)$ and differentiating gives the stated result. ■

These likelihoods can be mixed in the usual way (Proposition 3.2) to produce likelihoods with non-zero non-detection probabilities: $\tilde{p}_0 = p_0, \tilde{p}_1 = q_{01}p_0 + q_{11}p_1$. And by invoking independence between measurement lines we get a likelihood $\mathcal{P}_{\text{ord-stat}}(\mathbf{Z}|\mathbf{x})$ defined in the same way as $\mathcal{P}_{\text{gen}}$ (equation 3.3).

Note that the proposition requires the intersection innovation to occur precisely at the centre of the measurement line ($\nu = L/2$, or $\nu' = 0$). This is partially for convenience; one can still write down the likelihoods for other values of ν but they are rather messy piecewise functions. However, this special case is not as restrictive as it might seem: the case $\nu = L/2$ is particularly useful since, when the measurement lines are attached to the contours (see Section 3.2.3), every measurement line has $\nu = L/2$.

It was claimed at the start of this section that the order statistic likelihood would permit more general assumptions about the placement of clutter features: that is, more general than Assumption 4, which states the clutter features have a uniform distribution over the length of the measurement line. Although the proposition *does* adopt Assumption 4, it is easy to see how to adapt it for other background distributions, provided only that their cdfs are easy to calculate.

3.1.5 The contour likelihood ratio

An important advantage of the generative models described in this chapter is that they specify the likelihood of background clutter, as well as the likelihood for the target object. Thus one can calculate a likelihood ratio to test the hypothesis that the target is present against the null hypothesis that the measurements were generated by random background clutter. First we must specify exactly what is meant by the likelihood of background clutter.

Definition (Background likelihood) Fix a generative model for the image features $\mathbf{Z} = (\mathbf{z}^{(1)}, \dots \mathbf{z}^{(M)})$ detected on M measurement lines. Then the *background likelihood* $\mathcal{B}(\mathbf{Z})$ is just the likelihood of $\mathbf{Z}$ given that no target is present.

The fact that no target is present is equivalent to the intersection number of every measurement line being zero. So if, for example, the generative model is the generic one of Section 3.1.1, we can define the generic background likelihood by

$$\mathcal{B}_{\text{gen}}(\mathbf{Z}) = \prod_{i=1}^{M} p_0(\mathbf{z}^{(i)}). \tag{3.6}$$

Similar definitions apply for the other generative models discussed in sections 3.1.1–3.1.4, yielding $\mathcal{B}_{\text{poisson}}, \mathcal{B}_{\text{int-ext}}$ and $\mathcal{B}_{\text{ord-stat}}$.

Now we have formal expressions for the likelihood that a contour is present in a configuration $\mathbf{x}$, and the likelihood that it is not present at all. A natural step is to compare the two possibilities in a likelihood ratio. This is formalised in the definition of the *contour likelihood ratio.*

Definition (Contour likelihood ratio) Fix a generative model for the image features $\mathbf{Z} = (\mathbf{z}^{(1)}, \dots \mathbf{z}^{(M)})$ detected on M measurement lines. Let $\mathcal{P}(\mathbf{Z}|\mathbf{x})$ denote the likelihood for configuration $\mathbf{x} \in \mathcal{X}$ and $\mathcal{B}(\mathbf{Z})$ be the background likelihood. Then the *contour likelihood ratio* is defined to be

$$\mathcal{R}_{\mathbf{Z}}(\mathbf{x}) = \frac{\mathcal{P}(\mathbf{Z}|\mathbf{x})}{\mathcal{B}(\mathbf{Z})}. \tag{3.7}$$

When $\mathbf{Z}$ is clear from the context, we omit it and write $\mathcal{R}(\mathbf{x})$.

The contour likelihood ratio should be used with care, since the two models it compares (presence of target vs absence of target) have different numbers of parameters. In particular, the contour likelihood ratio should not be used for model selection. An information criterion such as AIC (Akaike, 1974) or one of its variants would be more appropriate, since these measures penalise models which have more free parameters.

3.1.6 Results and examples

Figure 3.4 gives examples of contour likelihood ratios on real images. The images themselves are shown in Figure 3.5. The target object, part of the boundary of a mouse, is assessed firstly in a plain background and then in a cluttered one. In each case, the log of the contour likelihood ratio is plotted for each of the four generative models discussed in this chapter.[4] The template, defined manually to be in the correct configuration, was offset by up to ± 20 pixels

[4]For the generic likelihood, the background probabilities $b(n)$ were set to $1/20$ for $n = 0, 1, \dots 19$ and zero otherwise. This is what the legend label "uniform background probabilities" refers to. See Section 3.2.1 for a discussion of this point.

in the x-direction only. For each offset value i, the image is measured obtaining the results $\mathbf{Z}_i$, so that the final plot is of

$$\log \mathcal{R}_{\mathbf{Z}_i}(\mathbf{x} \text{ shifted in } x\text{-direction by } i \text{ pixels}).$$

Some readers will be perplexed at the fact that measurements $\mathbf{Z}_i$ depend on the offset; this is the result of using measurement lines which are attached to the contour rather than fixed in the image. Section 3.2.3 describes why the dependence of $\mathbf{Z}$ on i is an acceptable approximation to having a single $\mathbf{Z}$ which is independent of the hypothesised configuration.

The differences between results for the various generative models reflect the assumptions which are made for these models. For instance, the interior-exterior likelihood produces the most confident interpretation of the template (that is, the highest likelihood ratio). This is because it employs the most specific information about the template — it incorporates knowledge of the occurrence of features in the interior of the mouse. On the other hand, the order statistic likelihood gives the least confident interpretation. This is because it uses only part of the measurements $\mathbf{Z}$ — namely, the most central feature detected on each measurement line, together with the number of features detected. With a plain background (Figure 3.4(a)), this turns out to be almost identical to the Poisson likelihood. A little thought reveals that this is correct: in an uncluttered environment, most measurement lines report only one feature (the target boundary), so the Poisson likelihood and order statistic likelihood end up making inferences on the same basis. In the cluttered environment (Figure 3.4b), the Poisson likelihood makes more complete use of the measurements and produces a more confident interpretation of $\mathbf{Z}$ than the order statistic likelihood.

Intuitive interpretations

Some of the expressions for contour likelihoods look a little messy.[5] In simple cases, however, the formulas can be given intuitive interpretations which have the flavour of combinatorics. Take the interior-exterior likelihood as an example, and consider a special case in which $q_{01} = 1$ (so the boundary of the target is never detected — inferences will be performed only on the distribution of interior and exterior features). Let the boundary density function $\mathcal{G}(z|\nu)$ be a Dirac delta function centred on ν (this corresponds to regarding the shape space as a perfect model for the target object and the edge detector as having no smoothing effects). Then it is easy to check that the likelihood $p_1^{\mathrm{ie}}(N; \mathbf{z}|\nu, d = \mathtt{False})$ given in Proposition 3.4 reduces to

$$p_1^{\mathrm{int-ext}}(N, \mathbf{z}) = \frac{1}{N!} \exp(-\nu(\mu - \lambda) - L\lambda)\mu^m \lambda^{N-m},$$

where m is the number of elements of $\mathbf{z}$ less than ν, μ is the density of the interior Poisson process and λ the density of the exterior or clutter process.

[5] Why should we care about this? One answer was given by G. H. Hardy (1940), who counselled "There is no permanent place in the world for ugly mathematics" in his *A Mathematician's Apology*.

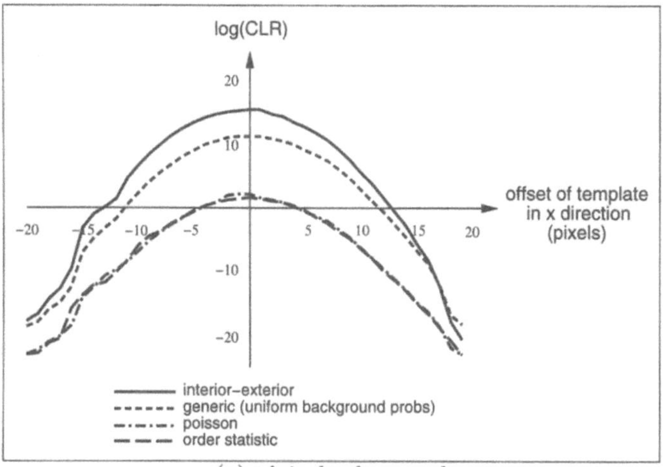

(a) plain background

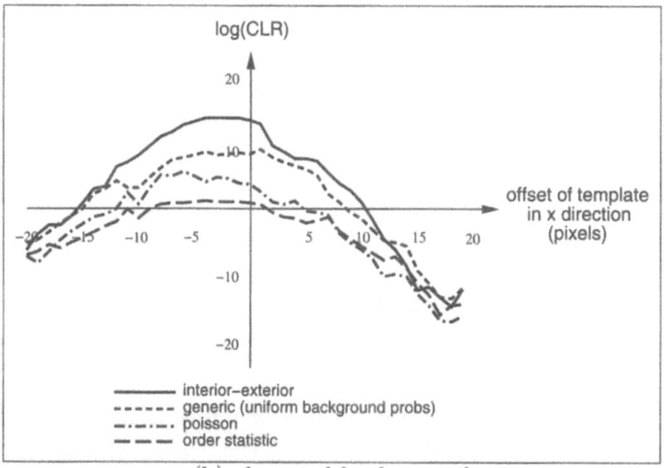

(b) cluttered background

Figure 3.4: ***Contour likelihoods.*** *A template placed on the mouse in each of the images in Figure 3.5 was offset in the x-direction only, and the various contour likelihoods calculated for each offset. (a) Results for the plain background (Figure 3.5(a)); (b) Results for the cluttered background (Figure 3.5(b)). Note that these are graphs of log-likelihood* ratios. *Hence the x-intercepts are positions where the measurements reflect the background clutter model and the target model equally well. The likelihood ratios for the cluttered case are more noisy than for the plain background, but have approximately the same shape (the systematic offset of the peaks is due to the high density of clutter features just to the left of the mouse). In particular, the x-intercepts which are the boundary between "target-like" and "clutter-like" regions remain the same (within a tolerance of around five pixels) for both the plain and cluttered backgrounds.*

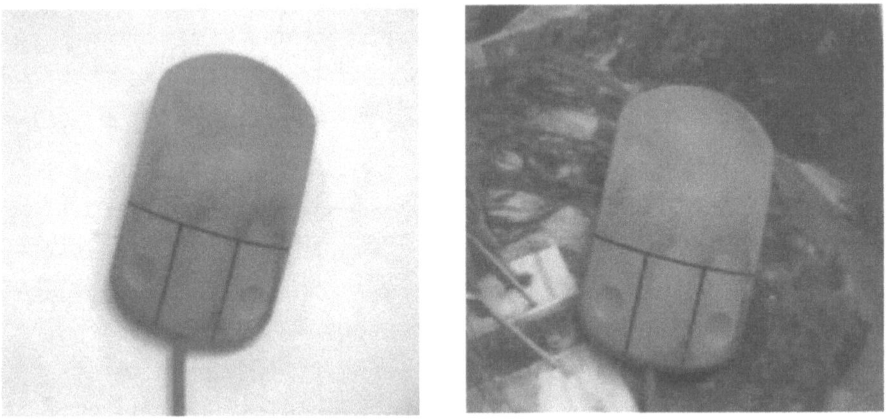

(a) plain (b) cluttered

Figure 3.5: ***Images used for comparison of contour likelihoods.***

The background likelihood remains the one given by Assumption 7:

$$p_0^{\text{int-ext}}(N, \mathbf{z}) = \frac{\exp(-L\lambda)\lambda^N}{N!}.$$

Taking the ratio of the last two expressions we find that

$$\frac{p_1^{\text{int-ext}}(N, \mathbf{z})}{p_0^{\text{int-ext}}(N, \mathbf{z})} = \exp(-\nu(\mu - \lambda))(\mu/\lambda)^m, \tag{3.8}$$

which is just the likelihood ratio for two one-dimensional Poisson processes with densities μ, λ on a length ν — precisely the interior-exterior model for the interior part of the measurement line. The exterior part of the line does not play a role in this likelihood ratio since the behaviour of features in that region is the same under the "clutter" and "target" hypotheses.

3.2 Background models and the selection of measurement lines

An advantage of the measurement line approach is that it reduces the problem of analysing a 2D image to that of analysing several 1D measurement lines. As will be argued in Section 3.2.2, the statistical processes generating features on different measurement lines can be treated as independent, which is why the generative models of the last section needed only to specify such processes on 1D subsets of the image.

3.2.1 Discussion of the background model

Recall that the numbers $b(n)$ specify the probability of obtaining n features on a measurement line positioned randomly on the background, and that these probabilities are learnt from typical training images. Of course this innocuous statement conceals a perennial problem in computer vision: how does one characterise a "typical" image, and even worse, how does one specify a prior for such images? Even when an image is reduced to the simple level of one-dimensional features, there is no straightforward answer to this question. However, it turns out the tracking and localisation systems described in this book are extremely robust to the choices of $b(n)$. Indeed, we routinely set $b(0) = b(1) = \ldots = b(n_{\max}) = 1/(1 + n_{\max})$ for some $n_{\max}$, with $b(n) = 0$ when $n > n_{\max}$. For measurement lines of 40 pixels, and an edge convolution operator with weights $(-0.375, -0.625, 0, 0.625, 0.375)$, one can take $n_{\max} = 20$ and obtain results indistinguishable from when the $b(n)$ are learnt from the entire sequence to be tracked. Another simple approach which gives equally good results in all our experiments is to learn the $b(n)$ from the first image of the video sequence to be tracked. Note that the numerical outputs can differ greatly for different choices of the $b(n)$: for instance, the only difference between the generic and Poisson results in Figure 3.4 lies in the choice of the $b(n)$. The robustness in tracking referred to above is an empirical fact which was observed despite these differences.

An alternative approach to modelling the occurrence of background features is the careful use of a Kalman filter framework to disregard spurious features (e.g. Peterfreund, 1998), but, in order for this to work in cluttered backgrounds, one needs much more accurate dynamical models than those available in the types of problem considered here. Other researchers explicitly adopt a uniform distribution on the $b(n)$ (e.g. Lowe, 1992), as suggested above.

Our second assumption about random background clutter features is that their *positions* are drawn from a uniform distribution. What is the corresponding assumption about 2D image features that would make this true? It would certainly hold provided the positions of all edgels of a given orientation were also distributed uniformly. We find this is sufficiently true over the small regions (scale around 40 pixels) occupied by the measurement lines, but it is clear that this approximation is unsatisfactory for larger regions. Further work is needed here: perhaps the recent ideas on filters and scale-invariance (Gidas and Mumford, 1999; Zhu et al., 1998) can be applied to obtain a more coherent theory.

3.2.2 Independence of measurement lines

The likelihood functions of this chapter were all derived by invoking Assumption 6, that feature occurrences on distinct measurement lines are statistically independent. Of course, this is only approximately true, since there are generally continuous edges in the background as well as on the boundary of the target object. There have been some attempts to allow explicitly for this type of dependence — for example, the Markov contour likelihood described later

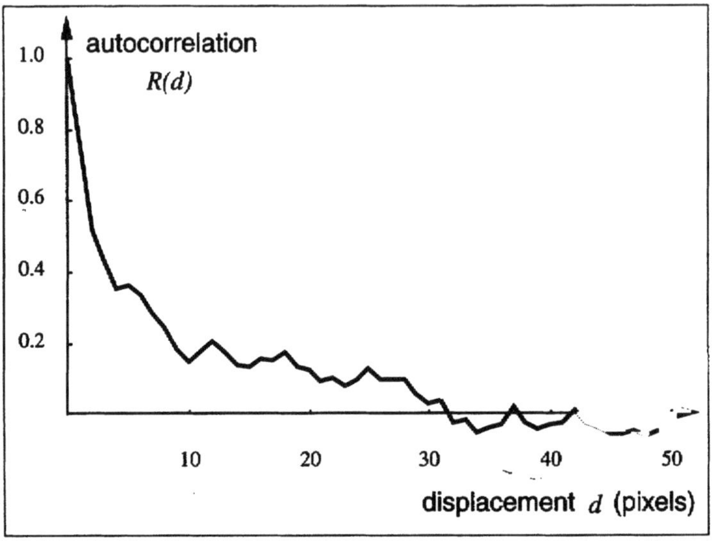

Figure 3.6: ***Feature autocorrelation is low for displacements of more than 30 pixels.*** *This is our justification for treating distinct measurement lines as statistically independent. The random process* $x(d)$ *whose autocorrelation is graphed here is described in the text and Figure 3.7, and the autocorrelation function is defined as usual by* $R(d) = (E[x(d)x(0)] - E[x(0)]^2)/(E[x(0)^2] - E[x(0)]^2)$.

in Chapter 5, or Markov random fields in general (Winkler, 1995). However, these are too computationally expensive for tracking tasks and in any case we find the independence approximation is acceptable if the measurement lines used for inferences are sufficiently far apart. A separation of 30 pixels is generally sufficient for the standard edge convolution operator described above, and this can be seen from Figure 3.6. This figure shows the autocorrelation of a random process $x(d)$ defined as follows (see also Figure 3.7). First, randomly position a measurement line, uniformly in position and orientation, on a typical background image. Apply a feature detector, select the closest feature to the centre of the measurement line, and define $x(0)$ to be the offset of this feature. The value of $x(d)$ is defined by first displacing the original measurement line a distance of d pixels in the direction of its normal, then applying the feature detector and setting $x(d)$ to be the offset of the most central feature.

As the graph shows, the autocorrelation of the most central feature on a measurement line drops to essentially zero after the measurement line has been displaced by 30 pixels. Of course, this does not prove the independence of the responses: even if the autocorrelation was precisely zero, it would only show that the moments $E(x(d)x(0))$ and $E(x(d))E(x(0))$ are equal. True statistical independence follows only if $E(x(d)^i x(0)^j) = E(x(d)^i)E(x(0)^j)$ for all positive i, j. Nevertheless, this empirical result is good evidence to justify our approach,

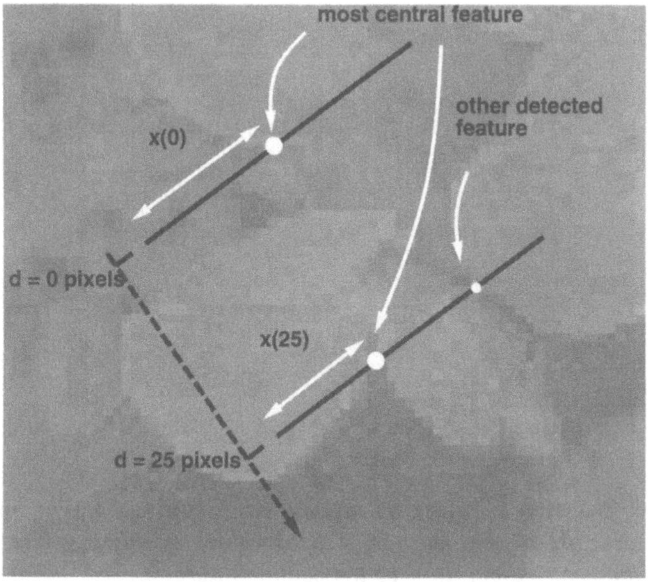

Figure 3.7: ***Investigating feature correlation.*** *The top solid black line is a measurement line positioned randomly on a typical background image. The value of the random process $x(d)$ is the offset of the most central detected feature after the initial measurement line has been displaced by d pixels in the direction of its normal.*

provided the measurement lines are separated by a normal distance of 30 or more pixels.

3.2.3 Selection of measurement lines

Often we need to perform Bayesian inference on the image, based on the measurements $\mathbf{Z}$ of several hypothesised configurations $\mathbf{x}_1, \ldots \mathbf{x}_n$. For Bayes' theorem to be valid, the set of measurement lines from which $\mathbf{Z}$ is obtained must be fixed in advance. However, it is sometimes convenient to allow the precise choice of measurement lines to depend on the configuration $\mathbf{x}$, as in Figure 3.8(a). This approach was often adopted in this book for ease of implementation. When the $\mathbf{x}_i$ are tightly clustered, the technique of allowing the measurement lines to vary with $\mathbf{x}$ turns out to be a close approximation to the statistically correct method of using fixed measurement lines.

A simple experiment was performed to provide evidence for this claim. Figure 3.8 shows the two regimes of measurement lines, termed "fixed" and "attached". The mouse-shaped template shown was offset by up to ± 20 pixels and the Poisson likelihood $\mathcal{P}_{\text{poisson}}$ calculated for each regime. The results are shown in Figure 3.9. For the fixed measurement line regime, the measurements

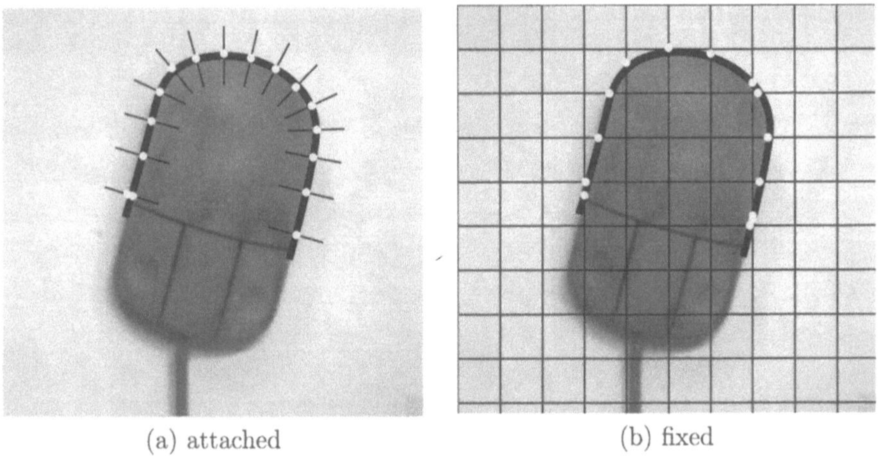

Figure 3.8: ***The two regimes of measurement lines.*** *(a) The measurement lines are attached to the contour at fixed values of the B-spline parameter. When a different configuration of the spline is hypothesised, the measurement lines will shift to have the same position relative to the new hypothesis. This means that any application of Bayes' rule is an approximation, but it has the advantage that the contour is known to intersect each measurement line at its mid-point, which simplifies the calculation of certain contour likelihoods. (b) The measurement lines are permanently fixed in the image. When a different configuration is hypothesised, no new measurements are performed, but instead a new set of intersections with the fixed measurement lines must be calculated. The fact that the intersection innovation is not known in advance makes some contour likelihoods harder to calculate.*

$\mathbf{Z}$ are made only once; the graph then consists of plotting

$$\text{constant} \times \mathcal{P}_{\text{poisson}}(\mathbf{Z}|\mathbf{x} \text{ shifted in } x\text{-direction by } i)$$

for each value of the offset i. The constant is chosen so that the area under the final graph is unity. Another way of thinking of this is that we have calculated the posterior based on an application of Bayes' rule with a uniform prior.

For the attached measurement line regime, things do not work out so cleanly. To use Bayes' rule, one should have a fixed set of measurements $\mathbf{Z}$, whereas in fact we now have a different set of measurements $\mathbf{Z}_i$ for each offset i. Nevertheless, we might hope that the *information* conveyed by each $\mathbf{Z}_i$ is roughly the same: this will certainly be true for small offsets i. In other words, the inferences performed using a given set of measurement lines yield almost identical results to inferences performed using the same set of measurement lines shifted by a few pixels. Hence, one could hope that plotting

$$\text{constant} \times \mathcal{P}_{\text{poisson}}(\mathbf{Z}_i|\mathbf{x} \text{ shifted in } x\text{-direction by } i)$$

(where the constant is calculated to give the graph unit area) would give similar

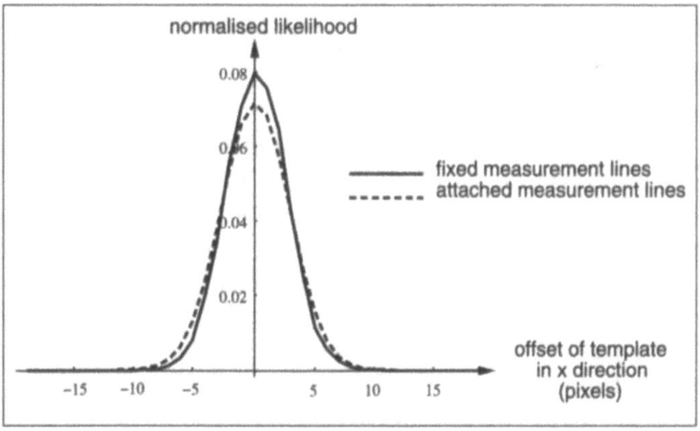

Figure 3.9: ***Posteriors calculated from the two regimes of measurement lines.*** *The posteriors are very similar, showing that the approximation caused by using the attached regime is an acceptable one. It is likely that the observable difference between the posteriors shown is due not to this approximation but the fact that the separation of measurement lines cannot be made exactly equal in the two regimes.*

results to the fixed measurement line regime. Figure 3.9 shows that this hope is justified: the posteriors produced by the two methods are very similar. In fact I suspect that the slight differences in the two graphs are not due to the "fixed vs attached" approximation, but rather because there is no convenient way to precisely enforce the same separation between measurement lines in the two cases. The grid of fixed measurement lines has a spacing of 30 pixels. The attached measurement lines have a spacing similar to this, but it varies slightly due to the parameterisation of the spline template, as can be seen from an examination of Figure 3.8(a).

Another interesting point to note in these graphs is that they are approximately Gaussian with an implied standard deviation of 4.5 pixels. The boundary density function[6] had a standard deviation of $\sigma = 7$ pixels for these calculations, but the inferences were aggregated over more than 15 different measurement lines, so one might expect the standard deviation to be reduced by a factor of around four. In fact several aspects of the model militate against this. Not all measurement lines are equally informative. For instance, those which are perpendicular to the direction of the offset do not reveal that the contour is misaligned. Also, the non-detection probability q_{01} ensures that each measurement line is not as "confident" in reducing the posterior variance as it would be if q_{01} were zero.

[6]Assumption 1, page 39.

3.3 A continuous analogue of the contour likelihood ratio

In the last section, it was argued that when the measurement lines are separated by a normal displacement of 30 pixels or more, the innovations of features detected on different measurement lines may be assumed to be independent. This, however, leaves open the question of what to do if the measurement lines are much closer together. Measurements on consecutive lines would then be correlated: is there a way to systematically take this dependence into account? Specifically, suppose there is a separation of τ between measurement lines, and we have likelihoods $\mathcal{P}_\tau \equiv \mathcal{P}_\tau(\mathbf{Z}_\tau|\mathbf{x})$, $\mathcal{B}_\tau \equiv \mathcal{B}_\tau(\mathbf{Z}_\tau|\mathbf{x})$, for the "contour" and "background" hypotheses respectively. (So $\mathbf{Z}_\tau$ is the set of measurements made on measurement lines of separation τ.) Then is there a way to take a meaningful limit as $\tau \to 0$? Maybe it is too much to ask that $\lim_{\tau\to 0} \mathcal{P}_\tau$, $\lim_{\tau\to 0} \mathcal{B}_\tau$ exist, but perhaps the likelihood ratio

$$\mathcal{R}_\tau = \frac{\mathcal{P}_\tau}{\mathcal{B}_\tau}$$

could make sense. This section proposes a model which appears to be statistically valid as $\tau \to 0$. The model has not been tested in rigorous experiments and is presented in the spirit of a useful idea which merits further investigation. The empirical results do not support our hope that $\mathcal{R}_\tau$ tends to a finite value as $\tau \to 0$, so it is clear that more work is required.

3.3.1 The continuous model

To present the theory in its most tractable form, several simplifying assumptions are made. First, we deal with the "attached" measurement line regime (Section 3.2.3), and allow only the *strongest* feature response on each line to be reported. This guarantees that precisely one innovation ν is detected corresponding to any particular value of the B-spline parameter s. That is, the measurements of a hypothesis $\mathbf{x}$ constitute a realisation $\nu_\mathbf{x}(s), s \in [0, T]$ of a random process — see Figure 3.10. The length T is just the number of spans in the B-spline, so that s traverses the whole spline. For convenience, innovations ν are measured from the centre of the measurement line, so that a feature detected on the hypothesised boundary $\mathbf{x}$ has an innovation of zero.

It is proposed to model $\nu(s)$ as a continuous auto-regressive process (ARP). As usual, we will form a likelihood ratio of two hypotheses. The hypothesis that the object is present in configuration $\mathbf{x}$ is abbreviated to H_O ("O" for "object"), and the hypothesis that no object is present is abbreviated to H_B ("B" for "background"). Under H_B, the ARP modelling $\nu(s)$ should essentially be a random walk on the length of the measurement lines, since there is no reason for the detected feature to be near the centre of the measurement line (see Figure 3.10). Under H_O, however, we expect $\nu(s)$ to remain near to the centre of the measurement lines, but to be perturbed by a certain amount of noise. These assumptions can be written down concretely as stochastic differential

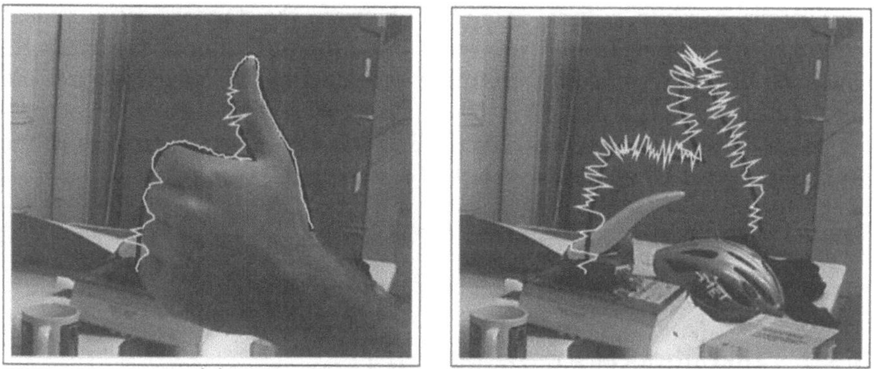

(a) (b)

Figure 3.10: ***Detected boundary as a continuous random process.*** *The thick black line is the hypothesised position* $\mathbf{x}$ *of the target, in this case a "thumbs-up" shape. The thin white line is the detected boundary, defined as the strongest feature on a measurement line. (a) The target is present at the configuration* $\mathbf{x}$*; this is the "object hypothesis"* H_O*. (b) The target object is not present; this is the "background hypothesis"* H_B*.*

equations (SDEs). Letting $w(s)$ denote a Wiener process with unit spectral density, the proposed model for each hypothesis is:

- under H_B:

$$\dot{\nu}(s) = h\dot{w}(s) \tag{3.9}$$

- under H_O:

$$\dot{\nu}(s) = -f\nu(s) + g\dot{w}(s). \tag{3.10}$$

Here a dot denotes differentiation with respect to s, and f, g, h are positive real constants to be set manually or learnt from training data. The values of g, h determine the amount of noise under H_B, H_O respectively, and the value of f determines how rapidly the process under H_O is attracted back to its (zero) mean position after a random deviation. If we think of s as time and ν as normal displacement from the contour, then the units of f are 1/time and the units of g, h are distance/time. Actually, equation 3.10 is the oldest example of a stochastic differential equation. It is known as the Ornstein-Uhlenbeck process (Uhlenbeck and Ornstein, 1930), and was considered as early as 1908 by Langevin (1908). It is clear that the random processes in Figure 3.10 do not have precisely the same properties as Ornstein-Uhlenbeck processes, but it turns out they are similar enough for some interesting and useful analysis. Some specific weaknesses in the model are discussed below. It is worth noting that similar SDEs have been used by Williams and Thornber (1999) and others for the problem of determining contour saliency, but there does not seem to be a direct connection between their work and the model proposed here.

There is, however, a direct connection between the standard measurement paradigm of Chapter 3, and the model just proposed. Recall the measurements $\mathbf{Z}_\tau$ of the hypothesis $\mathbf{x}$ are made according to the fixed measurement line regime, with a spacing of τ between the measurement lines. Then according to the model for $\nu(s)$ just described, $\mathbf{Z}_\tau$ is just a set of samples of the random process $\nu(s)$ at regular values of s: the samples are taken at

$$s_i = k\tau, \; k = 0, \dots M,$$

where $\tau = T/M$ and M is the number of measurement lines.

3.3.2 Likelihoods for H_O and H_B

The ARPs given by equations 3.9 and (3.10) are stationary Markov processes, so everything we need to know about them is summarised by their conditional probability densities after a time τ:

$$p(\nu(\tau)|\nu(0)) = \begin{cases} (2\pi h^2\tau)^{-1/2} \exp -\frac{1}{2}\frac{(\nu(\tau)-\nu(0))^2}{h^2\tau} & \text{under } H_B \\ \left(\frac{\pi g^2(1-e^{2f\tau})}{f}\right)^{-1/2} \exp -\frac{f(\nu(\tau)-e^{-f\tau}\nu(0))^2}{g^2(1-e^{-2f\tau})} & \text{under } H_O. \end{cases} \tag{3.11}$$

These expressions follow by integrating the SDEs given in equations 3.9 and 3.10. A simple treatment is given in Gelb (1974), and more detail can be found in Astrom and Wittenmark (1984), and Karatzas and Shreve (1988). Naturally, the density under H_B can be obtained from that for H_O by letting $g \to h$ and $f \to 0$, since when $f = 0$ and $g = h$ the two SDEs of equations 3.9 and 3.10 are the same.

Write $p_B(\nu)$ and $p_O(\nu)$ for the initial distribution of the background and object random processes respectively. Also write

$$\nu_0 = \nu(0), \nu_1 = \nu(\tau), \nu_2 = \nu(2\tau), \dots \; \nu_M = \nu(M\tau) = \nu(T).$$

Then by multiplying copies of equation 3.11 together, it's easy to see the log-likelihood, under H_B, of $\nu_0, \dots \nu_M$ is given by

$$l_B(\tau) = \log(p_B(\nu_0)) - \frac{M}{2}\log(2\pi h^2\tau) - \frac{1}{2h^2\tau}\sum_{i=1}^{M}(\nu_i - \nu_{i-1})^2. \tag{3.12}$$

Under H_O the log-likelihood is

$$\begin{aligned} l_O(\tau) = \log(p_O(\nu_0)) &- \frac{M}{2}\log\left(\frac{\pi g^2(1-e^{-2f\tau})}{f}\right) \\ &- \frac{f}{g^2(1-e^{-2f\tau})}\sum_{i=1}^{M}(\nu_i - e^{-f\tau}\nu_{i-1})^2. \end{aligned} \tag{3.13}$$

The log-likelihood ratio $\mathcal{L}$ is just the difference between equations 3.13 and 3.12:

$$\mathcal{L}(\tau) = l_O(\tau) - l_B(\tau). \tag{3.14}$$

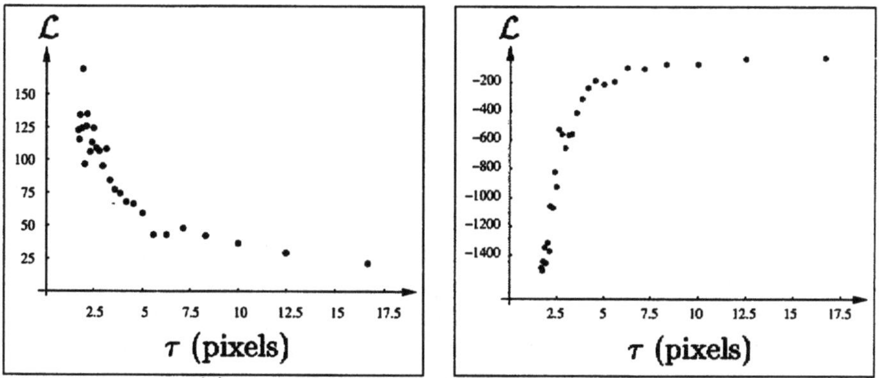

(a) (b)

Figure 3.11: ***Likelihoods for the "continuous" contour likelihood ratio.*** *Each graph shows a graph of $\mathcal{L}$, given by equation 3.14, for varying values of τ, the separation between measurement lines. (a) The graph for the image in Figure 3.10(a); notice that the value of $\mathcal{L}$ is positive for all values of τ. (b) The graph for the image in Figure 3.10(b); as expected, $\mathcal{L}$ is negative for this image. In both cases the likelihood ratio becomes more extreme for small values of τ.*

Figure 3.11 shows some results for the images in Figure 3.10. These results are intended to be a proof of concept only; more investigation would be required to establish the validity of the likelihood ratio given by equation 3.14 in practical applications. In particular, I was disappointed that $\mathcal{L}$ does not appear to tend to a finite limit as $\tau \to 0$. The parameters f, g, h were selected by trial and error, so that simulations of the ARP gave results looking more or less like Figure 3.10 (in fact we took $f = 0.1$, $g = 0.05$, $h = 0.2$). One important point to note is that regardless of the value of τ (corresponding to the horizontal axis in these graphs: τ is the separation of measurement lines in pixels), the log-likelihood $\mathcal{L}$ given by equation 3.14 has the correct sign. $\mathcal{L}$ is positive for the image containing the target, and negative for the control image containing only clutter.

3.3.3 Problems with the continuous ARP model

There are several clear weaknesses in using Ornstein-Uhlenbeck processes to model detected object boundaries. For one thing, the measurement process does *not* produce continuous outputs, even in the limit that the measurement lines are infinitesimally close: the detected feature which determines the value of ν can jump from one continuous edge to another (see, for example, the jump at the base of the thumb in Figure 3.10(a).) The use of a continuous model can be partially justified by considering that, in practice, the data will always be discretised at a known minimal separation — the pixel — and the signal could be continuously (and noisily!) interpolated between the data points. A related objection is that the "nearest feature" operation is non-linear, whereas the SDE is a linear model.

Another problem is that the ARPs as written do not account for the finite length of the measurement lines. The processes given by equations 3.9 and 3.10 are both free to wander far from the origin; in fact the variance of equation 3.9 grows without bound. Fortunately it is not difficult to define a process whose local properties are identical to equation 3.9 but which is constrained to stay within a finite length L. This uses the *reflection principle* described in Karatzas and Shreve (1988): we imagine the original process (equation 3.9) is reflected whenever it passes outside the interval $[-L/2, L/2]$. The conditional density function under H_B then becomes a sum of equation 3.11 over all points from which the unreflected path could have arrived:

$$p(\nu_n|\nu_0) = \text{const} \times \sum_{k=-\infty}^{\infty} \exp -\frac{1}{2}\frac{(\nu_n - 2kL \pm \nu_0)^2}{h^2\tau}, \tag{3.15}$$

where the constant is such that $p(\nu_n|\nu_0)$ integrates to one over $[-L/2, L/2]$. A similar sum gives the correctly constrained conditional density for the hypothesis H_O. The negative exponential means that few terms in the sum contribute: in practice, the measurement line length L is much greater than $h^2\tau$ and the three central terms $k = -1, 0, 1$ suffice for any values of ν.

4

Object localisation and tracking with contour likelihoods

Recall the main objective of the book: to infer the configurations of target objects in static scenes and in video sequences. The previous chapter introduced likelihood functions called contour likelihoods which can be used for such inferences. This chapter explains how to incorporate the contour likelihoods into algorithms for localisation and tracking. First, however, we pause to survey other approaches to localisation tasks.

4.1 A brief survey of object localisation

There is a vast literature on object localisation (and its more general cousin, object recognition), and in this section the main approaches used by vision researchers are summarised. Figure 4.1 attempts the rather impossible task of summarising this summary.

Algorithms for object recognition from a single static image must perform their search for the target in one of two types of space: the *pose space*, or a *correspondence space* (the terminology follows Grimson (1990)). The pose space is just the set of all possible configurations (denoted by $\mathcal{X}$ throughout this book). A correspondence space is the Cartesian product of the set of all features of a specified type (e.g. corners) together with the set of all features on the known model of the target. Note that the size of each space is exponential with respect to a crucial variable: for pose space, the size (i.e. volume) of the space is exponential in the dimension of the space, whereas for a correspondence space the number of possible matches is exponential in both the number of image features and the number of model features. This fact is perhaps the most fundamental reason that object recognition systems to rival humans are well beyond the reach of current vision research.

Nevertheless, it follows that when these crucial variables are sufficiently small, good progress can be made. Correlation techniques, for example, which correlate a known template with various regions in an image, can be run extremely fast in hardware but generally over a pose space of very small dimension. A relatively sophisticated example is Cootes et al. (1996), who used a statistical model of the grey-levels of every pixel in the target. A promising technique to have emerged recently is the Bayesian correlation of Sullivan et al. (1999), but it remains to be seen whether this will scale to high-dimensional problems.

The method proposed in this chapter, namely the factored sampling of contours, is also a pose space search, and certainly suffers the limitation that the dimension of this space must be small.[1] Indeed, experiments showed that if the prior is diffuse over more than three or four dimensions, the algorithm is too slow to be of practical use. But the method is still very flexible, and can be applied to pose spaces of say six to ten dimensions, provided the prior is tight over some of these. Other advantages are that the results have known confidence levels, that the method can be applied to a large class of targets with minimal modelling requirements, and that the target can be in a heavily cluttered environment.

One exception to the rule that the pose space must be low-dimensional is the class of Markov chain Monte Carlo (MCMC) methods — hardly surprising, as the early variants of this approach were originally popularised as the "solution" of exponential problems such as the travelling salesman. MCMC methods, when applied to object localisation (e.g. Grenander et al., 1991a; Mumford and Shah, 1985; Ripley and Sutherland, 1990), require a much less detailed model of the target than other approaches. However, these methods are particularly unsuited to general object localisation because they are not only slow, but require a good initialisation themselves for all but the simplest applications.

De Souza et al. (1997) have successfully combined an MCMC approach with a contour model whose framework is similar to the one outlined in this chapter. The targets in De Souza et al. (1997) are fish whose outlines are relatively complex, and a discussion comparing our approach with this work is given in section 4.2.1. Certain differences in the types of image analysed — in particular, the fact that the *strength* of edges is a useful indicator for the fish-tank images of De Souza et al. (1997) — mean the results are not directly comparable. The most important implication for this book is that MCMC methods can be applied to a contour framework.

An interesting method of guiding a search through pose space was proposed by Cameron and Durrant-Whyte (1988). In their scenario, obtaining measurements is very expensive (since they must be obtained by the laborious positioning of a robot arm), but computing the maxima of a function called the Probabilistic Membership Function over the pose space is computationally cheap. They used Bayesian reasoning to derive the optimal placement of

[1]Technically, the relevant quantity here is not the dimension of the pose space per se, but the ratio of the volume of the support of the prior to the support of the likelihood. The *survival rate* defined in Chapter 7 is a more reliable way of assessing the difficulty of a localisation problem.

		Advantages	Disadvantages	Examples
Pose space searches	correlation	• hardware readily available	• affected by lighting • can't deal with irregular shapes	Cootes and Taylor (1992, 1995); Cootes et al. (1996)
	MCMC	• deals with low SNR • target model can be very general	• needs very good initial guess • slow	Grenander et al. (1976–1981, 1991b); Green (1995); Miller et al. (1995); Shao and Mayhew (1996); Ripley and Sutherland (1990)
	factored sampling of contours	• known confidence levels • minimal target modelling required	• occasional false positives • target can't have complex 3D structure	this chapter
Correspondence space searches	Pruned search using one of: • geometric constraints • graph theory (cliques etc.) • relaxation labelling • invariants • probabilistic reasoning	• much faster than pose space search if pose space is high-dimensional	• target must have easily-detected features (e.g. corner, straight line, nose, mouth, particular invariants) • detailed modelling required	Lowe (1987a, 1991); Grimson and Huttenlocher (1990); Leung et al. (1995); Rothwell et al. (1992)
Other	Hough transform	• simple and effective when applicable	• limited effectiveness for shapes more complex than corners and straight lines	Duda and Hart (1973); Stockman et al. (1982)
	motion detection, optical flow and background subtraction	• little geometric modelling required	• needs moving target and still background	Koller et al. (1994)

Figure 4.1: ***Approaches to object localisation.***

successive sensor measurements.

The majority of the work done in object recognition has concentrated on searching a correspondence space. The astronomical number of matches (typically n^m with m, the number of model features, and n, the number of detected features, both in the range 100–1000) means the search must be pruned in some way. For example, if the chosen feature type is an invariant of some kind, the features can be labelled with parameters to guide the search. Other kinds of geometric and probabilistic reasoning can also dramatically reduce the number of possible matches. The obvious disadvantage of such an approach is that the target must possess easily-identified features of the type chosen (such as corners, or straight lines), and detailed models of the targets must be available. An approach in this vein which makes a particularly interesting comparison with the results in this chapter is Leung et al. (1995), which searched for faces in cluttered scenes using specific detectors for eyes, nose and the nose-lip junction followed by random graph matching.

The work of Lowe (1987b, 1992) is another example of an effective pruned search of feature-matching, in this case using oriented line segments. This work used some probabilistic arguments about the distribution of line segments to order a best-first search. Grimson and co-workers (1990, 1992) derived formal thresholds for the probability of random line segments and points matching more than a specified proportion of model features. Shimshoni and Ponce (1995) were able to rank hypothesised matches by probabilities, by calculating the distribution of certain length ratios and angles in 3D images of the targets.

The Hough transform is a well-known method for localising certain types of feature. In its most basic form, the Hough transform involves searching a correspondence space to create a function on the pose space whose maximum must be selected. However, much work (e.g. Duda and Hart, 1973; Stockman et al., 1982) has been devoted to highly efficient implementations of Hough transforms.

If we remove the restriction that only static images are used for localisation, other methods are available. Many systems (e.g. Koller et al., 1994) use motion detection, but of course this is only applicable when the target moves and the background is stationary. Another method, often combined with motion detection, is background subtraction (e.g. Soh et al., 1994). This requires a known background, and is surprisingly difficult to make robust to effects such as lighting changes and small amounts of camera motion.

4.2 Object localisation by factored sampling

The basic problem we would like to solve is: given a single image containing a single instance of the target object in the unknown configuration $\mathbf{x}$, estimate $\mathbf{x}$. A harder problem which we would also like to solve is: given a single image containing zero or more target objects in unknown configurations, estimate the number of targets present and their configurations. The second problem is tackled in Section 4.3; for the moment let us concentrate on the first problem.

In classical estimation theory (Papoulis, 1990), an estimator $\hat{\mathbf{x}} : \mathcal{Z} \to \mathcal{X}$ mapping the measurements $\mathbf{Z}$ to an estimate $\hat{\mathbf{x}}(\mathbf{Z})$ would be defined. Standard choices for $\hat{\mathbf{x}}$ would be

- **Maximum likelihood estimate (MLE)**: set

$$\hat{\mathbf{x}} = \arg\max_{\mathbf{x}} \mathcal{P}(\mathbf{Z}|\mathbf{x}),$$

  where $\mathcal{P}(\mathbf{Z}|\mathbf{x})$ is the likelihood of the measurements $\mathbf{Z}$.

- **Minimum mean squared error (MMSE)**: given a prior $p(\mathbf{x})$, calculate the posterior $p'(\mathbf{x}|\mathbf{Z})$ using Bayes' theorem ($p' \propto \mathcal{P}p$) and then set

$$\hat{\mathbf{x}} = \mathbb{E}_{p'}\,\mathbf{x},$$

  where $\mathbb{E}_{p'}$ denotes expectation with respect to $p'(\mathbf{x}|\mathbf{Z})$.

- **Maximum a posteriori (MAP) estimate** :[2] Given a prior $p(\mathbf{x})$, calculate the posterior p' as before and set

$$\hat{\mathbf{x}} = \arg\max_{\mathbf{x}} p'(\mathbf{x}|\mathbf{Z}) = \arg\max_{\mathbf{x}} \mathcal{P}(\mathbf{Z}|\mathbf{x})p(\mathbf{x}).$$

The method proposed here does not directly solve any of these problems: it is a stochastic, approximate solution which can be adapted so that its output converges to either the MLE or the MMSE, as the amount of computation permitted tends to infinity. It can also be used to obtain MAP estimates but is not particularly well suited for this. The method is called *factored sampling* and requires three inputs: the number of samples n, the sampling distribution $q(\mathbf{x})\,d\mathbf{x}$, and the likelihood $\mathcal{P}(\mathbf{Z}|\mathbf{x})$. The basic idea is to draw random samples from $q(\mathbf{x})$ and evaluate the likelihood of each. More formally, we have the following.

Definition (Localisation by factored sampling) Given inputs $q, \mathcal{P}$ and the configuration space $\mathcal{X}$, *localisation by factored sampling* is a stochastic algorithm which outputs a particle set $(\mathbf{x}_i, \pi_i)_{i=1}^n$, $n = 1, 2, \ldots$ defined by

$$\mathbf{x}_i^{(n)} \vdash q(\mathbf{x})$$
$$\pi_i^{(n)} = \frac{\mathcal{P}(\mathbf{Z}|\mathbf{x}_i^{(n)})}{\sum_{j=1}^n \mathcal{P}(\mathbf{Z}|\mathbf{x}_j^{(n)})}$$

(see page 13 for the definition of $\vdash$). How can factored sampling be used for object localisation? The next proposition tells us the answer: for an MLE, use any sampling distribution q and take $\hat{\mathbf{x}}$ to be the particle with the highest weight; for the MMSE, use the prior p as the sampling distribution and take $\hat{\mathbf{x}}$ to be the mean of the $\mathbf{x}_i$ weighted by the π_i.

---

[2]The arguments about whether MAP estimates are useful or meaningful are well trodden, and will not be resolved here! MAP estimates are not used in this book except for the discrete distribution over the number of targets in Section 4.3.

Proposition 4.1 *Let $(\mathbf{x}_i, \pi_i)_{i=1}^n$ be the particle set produced by factored sampling with sampling distribution $q(\mathbf{x})$ and likelihood $\mathcal{P}(\mathbf{Z}|\mathbf{x})$. Assume the MLE and MMSE are unique. Then:*

(a) Let

$$\hat{\mathbf{x}}^{(n)} = \text{ the } \mathbf{x}_i^{(n)} \text{ with highest weight } \pi_i^{(n)},\ i = 1, \dots n.$$

Then $\hat{\mathbf{x}}^{(n)}$ tends (almost surely) to the MLE as $n \to \infty$.

(b) Let

$$\overline{\mathbf{x}}^{(n)} = \sum_{i=1}^{n} \pi_i^{(n)} \mathbf{x}_i^{(n)}.$$

If $q(\mathbf{x}) = p(\mathbf{x})$, the prior, then $\overline{\mathbf{x}}^{(n)}$ tends (almost surely) to the MMSE as $n \to \infty$.

Moreover, (a) still holds if every $\mathbf{x}_i$ is replaced by some $\mathbf{x}'_i$ with a higher likelihood.

Remarks. The last statement is much more general than we need: the idea is that we want to speed convergence by replacing each $\mathbf{x}_i$ by $\arg\max_{\mathbf{x}} \mathcal{P}(\mathbf{Z}|\mathbf{x})$, where the maximisation takes place over a small neighbourhood of $\mathbf{x}_i$.

Proof. (a) An easy exercise: because q is positive we are guaranteed that particles get arbitrarily close to the MLE as n becomes large.

(b) This follows immediately from the factored sampling theorem of Grenander, which is restated as Theorem 2.6 in Section 2.5. The last statement is another easy exercise. ∎

Rigorously stated propositions can sometimes conceal what is really going on, and this is a case in point. Often the likelihood $\mathcal{P}$ is very sharply peaked; in this case the random sampling is more like a random search for the narrow peak. Rather than convergence, we are more concerned with whether or not there will be *any* particles inside the peak. Section 4.5 discusses this possibility in more detail. More generally, note that we have stated nothing about the speed of convergence: to be of any use, the factored sampling method of object localisation must achieve acceptable convergence at moderate values of n. Moreover it must be possible to choose a sensible sampling distribution q and a realistic likelihood $\mathcal{P}$. For the likelihood, one can use any of the contour likelihoods introduced in Chapter 3, and Section 4.4 describes how to learn a suitable prior $p(\mathbf{x})$, which can also be used as the sampling distribution $q(\mathbf{x})$. But now let us move swiftly on to an empirical "proof" that the convergence in the last proposition really is fast enough for our purposes: some successful experimental results.

4.2.1 Results

Maximum likelihood estimates

Consider the task of approximating the MLE; the first two examples locate a mouse and a cyclist respectively. First a contour template was obtained manually (Figure 4.2(a) and (d)). A generic sampling distribution $q(\mathbf{x})$ was specified: uniform over x and y translation, scaling by a factor drawn from $N(1, 0.08^2)$, together with some rotation for the cyclist (standard deviation of a few degrees) and affine deformation for the mouse. The reason we need not bother with careful modelling of q is that Proposition 4.1 tells us that *any* strictly positive sampling distribution will do.

Factored sampling was applied to different scenes (Figure 4.2(b) and (e)), with $n = 1000$ particles, which were evaluated by the interior-exterior likelihood $\mathcal{P}_{\text{int-ext}}$ (equation 3.5). As permitted by the last part of Proposition 4.1, each particle was replaced with a more likely one: in fact the regularised least-squares fit described in Blake and Isard (1998) was applied to each particle before its likelihood was calculated. Figure 4.3 gives details of the parameters used. The procedure takes 1.7 seconds on an SGI O2 (R5000, 180 MHz). Figure 4.2(c) and (f) shows the results: in both cases the output $\hat{\mathbf{x}}$ was an acceptable approximation to our human assessment of what the MLE should be.

Although the MLE procedure is only really defined for images containing exactly one target object, it is interesting to observe the behaviour for a scene with more than one. So the same experiment was tried with a head-and-shoulders template (Figure 4.4) in an office scene. The sampling distribution for this experiment was learnt by the method described later, in Section 4.4. The results are shown in Figure 4.5, where we have shown the ten most likely configurations $\mathbf{x}$ from the set of 1000 particles. They are ranked by their contour likelihood ratio $\mathcal{R}$. Note the values of $\mathcal{R}$ are all greater than one (indicating the hypothesis $\mathbf{x}$ being more likely than the hypothesis of random background clutter).

It is natural to wonder whether the complexity of the interior-exterior generative model is necessary for these applications, especially if there are more easily calculated alternatives which perform just as well. As an example, consider the Poisson likelihood $\mathcal{P}_{\text{poisson}}$ (equation 3.4). It was noted earlier that this is equivalent to the likelihood used by Isard and Blake (1996) for evaluating particle sets in their original Condensation tracker.

The same particle set as before was ranked using the likelihood $\mathcal{P}_{\text{poisson}}$, and the result is shown in Figure 4.6. The performance of this likelihood was worse than the interior-exterior likelihood, in the sense that more spurious hypotheses are ranked above the true target configurations. The modelling of interior features has brought two benefits:

- $\mathcal{P}_{\text{int-ext}}$ takes into account the model of how many interior features are expected on each measurement line.

- $\mathcal{P}_{\text{int-ext}}$ gives higher probabilities to lines with an unusual partitioning of features into interior and exterior — this can be seen in the m-dependence of (equation 3.8), for example.

Both of these differences contribute to the fact that $\mathcal{P}_{\text{poisson}}$ is easily distracted by the bookcase region (see Figure 4.6), in which the density of clutter features is very high.

(a) mouse template (d) cyclist template

(b) scene to examine (e) scene to examine

(c) localisation output (f) localisation output

Figure 4.2: ***Localisation results.*** *In each example, 1000 particles were drawn from the prior. In (c) and (f), the particle of highest contour likelihood ratio is shown in black, and the second and third highest are in white. (In (c), the three contours are nearly identical.) Both priors are uniform over the two translation parameters; the mouse prior is the same as for Figure 4.5, and the cycle prior is also the same except that it supports much less 2D rotation. In both cases, the background and lighting are different in the template and examined scene.*

Non-detection probability	q_{01}	0.05 for mouse, 0.2 for cyclist
Clutter feature probabilities	$b(n)$	0.05 for $n = 0, 1, \ldots 19$ and 0 otherwise
Interior model parameters		MLE of Poisson densities from template
Boundary feature distribution	$\mathcal{G}(z\|\nu)$	Gaussian with std dev of 7 pixels
Length of measurement lines	L	40 pixels
Prior	$p(\mathbf{x})$	described in the text

Figure 4.3: ***Parameter values and other choices used for experiments.*** *The non-detection probabilities were set manually after observing the templates for each object. The interior-exterior likelihood $\mathcal{P}_{\text{int-ext}}$ (Section 3.1.3) was used to evaluate the particles, with the "attached" measurement line regime (Section 3.2.3).*

Minimum mean square error estimate

Very similar results are obtained from the MMSE approach. In the case of localising the mouse and the cyclist (Figure 4.2), the MMSE is indistinguishable from the MLE (we used the $q(\mathbf{x})$ defined earlier as a generic prior). A more interesting case is the office scene. Taking $n = 10,000$ but with the other parameters as before, factored sampling gives the output shown in Figure 4.7. Since more than one target is present, it does not really make sense to take the mean of this distribution. Instead, it is better to regard the output as a particle representation of the posterior distribution, as guaranteed by the factored sampling theorem (Theorem 2.6, page 25).

4.2.2 Interpretation of the gradient threshold

As mentioned in the survey of localisation techniques (section 4.1), some work by de Souza et al. (1997) has combined an MCMC searching method with a contour framework involving the detection of features on measurement lines. In their approach, precisely one feature is detected on each measurement line. The location of this feature is determined by a maximum likelihood estimate for the innovation; a key difference to our approach is that the *strength* of the potential feature is incorporated into the estimate. They set

$$\hat{z} = \arg\max s(z) \exp(-\frac{(z)^2}{2\sigma^2}),$$

where $s(z)$ is the strength of the output of the chosen feature detector.

The assumption essential for a mathematical justification of such an MLE is that the intensity of a single ideal feature (e.g. a step edge) is perturbed by additive Gaussian noise. Therefore, incorporating feature strength is certainly

Figure 4.4: ***Template and interior model.*** *This image was used to obtain a head-and-shoulders template, and to learn the Poisson densities of interior features. This template and model were used to produce the results of Figure 4.5.*

a good idea in the absence of clutter, and is a good approximation when the strength of any clutter features is expected to be less than that of the target feature. This would appear to be the case in the fish-tank images examined in De Souza et al. (1997). However, what happens when we extend this MLE approach to a more complex model in which an unknown number of clutter features are present, whose strengths are unknown but comparable to (and perhaps greater than) the target's? This is precisely the problem of identifying change-points in a step function! The change-point problem has been studied for many years by statisticians, and the solutions (perhaps the most recent is Green (1995)) are computationally expensive, requiring several seconds of CPU time at best. This is clearly inappropriate for the application envisaged here, where the change-points must be determined on each measurement line in each of many configurations.

Hence the approach of Chapter 3: *define* what we mean by a feature, *model* the occurrence of these features with simple distributions, and *infer* results based on the model. In our case, to define a feature we must first specify an operator which will be applied to the measurement line, and then select a threshold called the *gradient threshold*; outputs of the operator which are above the gradient threshold are regarded as features. Note in particular that the entire probabilistic framework described in Chapter 3 is valid for any choice of the gradient threshold. However, the inferences which can be made will be

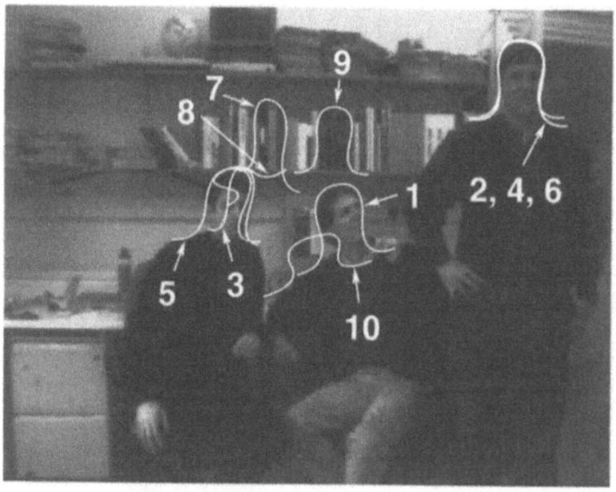

Ranking	Description	$\mathcal{R}(\mathbf{x})$
1	middle target	408.1
2	right target	58.6
3	left target*	34.1
4	right target	25.1
5	left target	14.9
6	right target	8.4
7	spurious	6.3
8	spurious	4.5
9	spurious	2.4
10	spurious	1.9

*partial hit

Figure 4.5: ***Finding head-and-shoulders outlines in an office scene using the interior-exterior likelihood.*** *The results for a particle set of* 1000 *configurations are shown ranked by their likelihood. The table shows the numerical values of the contour likelihood ratio* $\mathcal{R}$ *(equation 3.7, page 51); a value greater than one means a configuration is more target-like than clutter-like.*

more useful (i.e. discriminating) when the threshold is such that the number of clutter features detected is low, and the number of target boundary features detected is high.

To see this more clearly, consider limiting cases of the gradient threshold. If it is very high, then no features will be detected; the contour likelihood ratio can still be calculated but is equal to one everywhere. If the gradient threshold is very low, many features will be detected. However, it is still conceivable that useful inferences could be made, if the distribution of internal features is sufficiently different to the clutter — recall the simplified example at the end of section 3.1.6, where a useful inference could be drawn from the number of features detected on the interior portion of the measurement line.

It is intuitively (and experimentally) clear that for most targets there is an optimum threshold between these two extremes. For the results in this section we made no mathematical attempt to find this optimum; the gradient threshold was adjusted to the highest value for which most of the boundary features are detected. This then determines the non-detection parameter q_{01} — if the "most" of the previous sentence was 90%, then q was set to 0.1.

4.3 Estimating the number of targets

So far, the object localisation problem has been restricted to the case where it is known there is exactly one target object present in an image. Can we make any progress if the number of targets is not known in advance? The answer

Figure 4.6: ***Worse performance with a simpler likelihood.*** *The best ten of the same 1000 particles as in Figure 4.5 are shown, this time ranked by the simpler likelihood* $\mathcal{P}_{\text{poisson}}$. *The performance is worse, showing that adopting a separate model for the interior features is beneficial.*

is yes; this section explains the theory behind the multiple target recognition problem and describes a practical implementation.

Throughout this section, k will be the number of targets present. If the configuration space for a single target is $\mathcal{X}$, then the configuration space for the multiple target problem is

$$\widetilde{\mathcal{X}} = \bigcup_{k=0}^{\infty} \mathcal{X}^k.$$

A typical element of this space is written as $\widetilde{\mathbf{x}} = \{k; \mathbf{x}_1, \dots \mathbf{x}_k\}$ where each of the $\mathbf{x}_j \in \mathcal{X}$. Formally, it is easy enough to write down a likelihood for this space. As usual, let $\mathbf{Z} = (\mathbf{z}^{(1)}, \dots \mathbf{z}^{(M)})$ be the features detected on the M measurement lines in an image. The definitions of intersection number and intersection innovations (Section 3.1.1) generalise in the obvious way to the multiple target situation. So $c_{\widetilde{\mathbf{x}}}(i)$ is just the number of points at which the ith measurement line intersects any of the contours in the configuration $\widetilde{\mathbf{x}}$. Similarly $\boldsymbol{\nu}_{\widetilde{\mathbf{x}}}(i)$ is the set of distances of these points from the start of the measurement line. Assume we have adopted one of the generative models of Chapter 3, and can therefore calculate the likelihood function $\tilde{p}_{c_{\widetilde{\mathbf{x}}}(i)}(\mathbf{z}|\boldsymbol{\nu}_{\widetilde{\mathbf{x}}}(i))$ for the ith measurement line. Then assuming (as usual) independence between

Figure 4.7: ***Factored sampling on an office scene using the interior-exterior likelihood.*** *The output is a particle representation of the posterior distribution. Particles are shown with mass proportional to their weights; there are discernible peaks in the distribution at each of the three targets.*

measurement lines, the multiple target likelihood is just

$$\mathcal{P}(\mathbf{Z}|\widetilde{\mathbf{x}}) = \prod_{i=1}^{M} \tilde{p}_{c_{\widetilde{\mathbf{x}}}(i)}(\mathbf{z}^{(i)}|\boldsymbol{\nu}_{\widetilde{\mathbf{x}}}(i)),$$

exactly the same as equation 3.3 except that $\mathbf{x} \in \mathcal{X}$ has been replaced by $\widetilde{\mathbf{x}} \in \widetilde{\mathcal{X}}$.

For any fixed k, multiple target object location can in principle proceed just as in the single target case, by factored sampling: draw samples $\widetilde{\mathbf{x}}_i$ from a prior $p(\widetilde{\mathbf{x}})$ on $\widetilde{\mathcal{X}}$, and calculate approximations to either the MLE or MMSE as in Proposition 4.1. Also, one can easily obtain estimates of the posterior probability for k (the number of targets) just by summing the weights of all the samples with a given value of k.

In practice, it is necessary to make some approximations to avoid a combinatorial explosion of the computation. Figure 4.8 gives an example. It assumes a Poisson prior on k, with mean one. The prior $p(\mathbf{x})$ for a single target is uniform in x and y translation and permits a small amount of affine deformation. The prior $p(\widetilde{\mathbf{x}}|k)$ for k targets consists of k independent copies of $p(\mathbf{x})$, except that the distance[3] between every pair of targets must be greater than a threshold

[3] In the spline space metric on $\mathcal{X}$ — see Appendix A.1.

Number of targets:	0	1	2	3	4
Prior probabilities:	0.36	0.36	0.18	0.06	0.02
Posterior probabilities:	0.01	0.14	0.71	0.13	0.00

Figure 4.8: ***Estimating posterior probabilities for the number of targets.*** *The MAP estimate for the number of targets is two, which agrees with a human assessment of the number of coffee mugs in this picture. The posterior probabilities are estimated by sampling from the prior $p(\widetilde{\mathbf{x}})$ described in the text (which does not allow distinct targets to be too close to each other). However, to reduce the computational requirements, only the eight configurations shown are used for the estimation.*

τ:

$$p(\widetilde{\mathbf{x}}|k) \propto \begin{cases} \prod_{i=1}^{k} p(\mathbf{x}_i) & \text{if } d(\mathbf{x}_i, \mathbf{x}_j) > \tau \text{ for all } i \neq j;\ i, j = 1, \ldots k \\ 0 & \text{otherwise.} \end{cases} \tag{4.1}$$

For this example we took $\tau = 40$ pixels — the length of the measurement lines. This minimum-distance restriction was introduced for convenience; it ensures the intersection numbers for the measurement lines are almost all zero or one, so the likelihoods of Chapter 3 can be used. (The more sophisticated multi-target likelihoods of Chapter 6 would remove the need for this restriction, since they allow intersection numbers $c \geq 2$. If you like, (equation 4.1) is just a naïve version of the exclusion principle to be introduced in Chapter 6.) Even with this restriction, the space $\widetilde{\mathcal{X}}$ is too large to be explored by straightforwardly sampling from some prior $p(\widetilde{\mathbf{x}})$. Layered sampling (Sullivan et al., 1999) and partitioned sampling (Chapter 7) would both be applicable here, but a simpler method was shown to produce satisfactory results. First, 10,000 contours were drawn from the single target prior $p(\mathbf{x})$ described above. Their single-target likelihoods were calculated and the eight contours of highest likelihood which mutually satisfied the distance restriction were retained. (That is, set $\mathbf{x}_1$ to be the contour with highest likelihood; then set $\mathbf{x}_2$ to be the highest remaining contour such that $d(\mathbf{x}_1, \mathbf{x}_2) > \tau$, and so on.) Finally, the set of all k-tuples of the eight contours was used to evaluate the posterior probability for k targets being present. Only the values $k = 0, 1, 2, 3, 4$ were considered.

As can be seen from Figure 4.8, this method correctly infers the number of coffee mugs present, yielding a MAP estimate of two.

4.4 Learning the prior

An essential ingredient in the factored sampling method is a prescription for choosing either a prior $p(\mathbf{x})$ on $\mathcal{X}$. The construction of $p(\mathbf{x})$ depends both on the intended application, and the degree of generality expected from the system.

If we have no information about the expected configurations, the most natural choice may be a uniform density on $\mathcal{X}$. (The concept of a uniform density on $\mathcal{X}$ can be made precise, and details of how to do this are given in Appendix A.1.) Another method found to be very successful in practice is to learn the prior from a video sequence, as follows.

1. Obtain a video sequence of the target object undergoing typical motion. The requirement on the length of the video sequence is that the object should sweep out its configuration space. (The definition of "sweep out" depends on the way the data will be processed. If, for example, $\mathcal{X}$ is divided into discrete bins, then we would say the sequence has swept out $\mathcal{X}$ if every bin with non-negligible prior probability contains several data points.) For six-dimensional $\mathcal{X}$, the implementation used in this chapter found that sequences of 30 seconds were ample.

2. Initialise a contour tracker by hand, and track the object throughout the sequence. Sequences for this chapter were tracked using a Kalman filter contour tracker (Blake et al., 1993; Terzopoulos and Szeliski, 1992).

3. Convert the resulting data into a distribution on $\mathcal{X}$, and smooth if necessary. In this chapter, the distribution was calculated by histogram, and no smoothing was necessary.

4. "Redistribute" the resulting distribution to take account of any known priors on sub-manifolds of $\mathcal{X}$, and of known relationships between sub-manifolds. As a very simple example suppose $\mathcal{X}$ comprises 2D translations on the unit square, and 2D rotations. In mathematical notation, $\mathcal{X} = I^2 \times SO(2)$, where I is the unit interval. Let the density obtained after step 4 be $g(\mathbf{x})$, and suppose we know that, despite any information to the contrary in the training sequence, all orientations of the target object are equally likely. Then to obtain a more correct prior $p(\mathbf{x})$ from $g(\mathbf{x})$, we should redistribute the values of g around all cosets of $SO(2)$ by setting

$$p(\mathbf{x}) = \int_{\mathbf{x}' \in [\mathbf{x}]} g(\mathbf{x}') d\mu.$$

Here $[\mathbf{x}]$ denotes the coset of $SO(2)$ containing $\mathbf{x}$ (i.e. the set of all configurations created by 2D rotations of $\mathbf{x}$) and μ is the uniform measure on $SO(2)$.

Or perhaps we know that, despite any information to the contrary in the training sequence, the target object's prior orientation is independent of its prior translation. In this case we need to write $\mathbf{x} = (\mathbf{t}, \theta)$, where $\mathbf{t}$ is a 2D translation and θ the angle of rotation. Now set $p(\mathbf{x}) = u(\mathbf{t})v(\theta)$, where

$$u(\mathbf{t}) = \int_{\mathbf{x}=(\mathbf{t},\cdot)} g(\mathbf{x})d\mu,$$

and

$$v(\theta) = \int_{\mathbf{x}=(\cdot,\theta)} g(\mathbf{x})d\lambda.$$

Here λ denotes the uniform measure on the unit-square.

The implementation described here includes switches for enforcing uniformity over translations, 2D rotations, or 3D rotations, and for enforcing independence between translations and all other parameters.

Note there is actually no need for the training sequence to resemble the test sequences in every respect: the only requirement is that the observed values of $\mathcal{X}$ are representative of the prior. Therefore, a sequence without clutter, or with a distinctively coloured target object can be used.

The result of sampling from a prior created like this is shown in figure 4.9.

4.5 Random sampling: some traps for the unwary

This section presents a mathematical analysis of an idealised factored sampling scenario. The objective is to impart the flavour of the localisation algorithm presented earlier, in a much simpler context where the analysis is transparent. The lesson to be learnt from this section is as follows: when searching for the maximum of a given function by random sampling, one should sample from a distribution which is *broader* than one's belief about the distribution of where the maximum may lie.

To demonstrate this lesson, a precise formulation of the problem is needed. Let X be a continuous random variable with pdf $f(x)$. X is the *target* variable — we sample from it once, obtaining say x_0. In this scenario, $f(x)$ is known to the localisation algorithm but x_0 is not. The localisation algorithm will try to guess a value sufficiently close to x_0 by sampling from another distribution. The pdf to sample from is denoted $g(y)$, and write $Y_1, Y_2, \ldots$ for a sequence of i.i.d. random variables with this distribution.

Recall the discussion following Proposition 4.1: often the likelihood $\mathcal{P}$ is so sharply peaked that we are more concerned with whether or not any particles fall in the peak than with the convergence of the factored sampling. With this

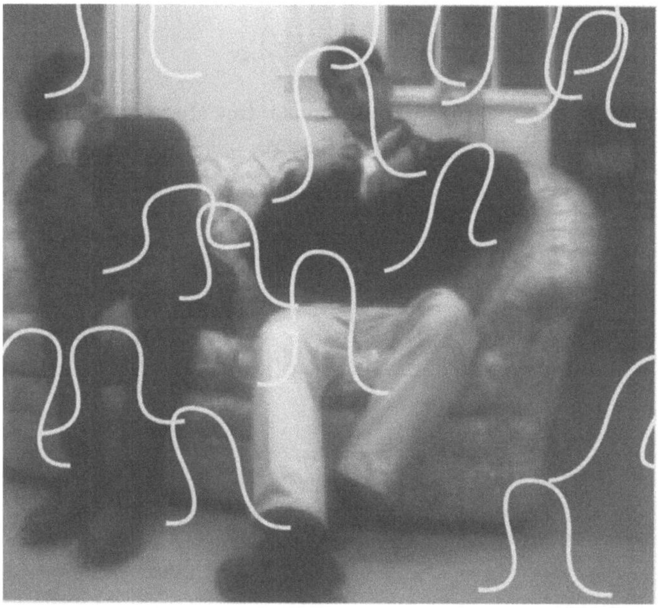

Figure 4.9: ***The prior density.*** *Shown here are 15 samples from a prior for head-and-shoulders outlines constructed by the learning method described in the text.* $\mathcal{X}$ *is the six-dimensional space of affine transformations. This prior is the one used to produce the results of Figure 4.5. The learnt motions comprised the head tilting up to about* 40° *from the vertical in every direction, and the prior was redistributed afterwards to have the following properties: uniform over both translation parameters, and a Euclidean scaling factor obeying a normal distribution with a mean of one and standard deviation about 7%.*

as motivation, fix a small real number $\tau > 0$ (called the *hit threshold*) and call Y_n a *hit* if $|y_n - x_0| < \tau/2$. Finally, define a discrete random variable H, called the *hit time*, by setting

$$H = \text{ least } n \text{ such that } Y_n \text{ is a hit.}$$

The main questions are (a) what is EH? and (b) what choice of $g(y)$ minimises EH? Most people's first guess for g is to set $g = f$, but, as we show below, this is certainly not the optimal choice, and in some cases (such as when f is a normal distribution) it leads to disasters such as $EH = \infty$! In fact, the solution is to take $g \propto \sqrt{f}$, but this is not always a sensible choice since we may also want $\text{var}(H)$ to be small.

Let $\alpha(x) = \text{Prob}(|y - x| < \tau/2)$. Then a simple calculation shows that

$$E(H|X = x) = \frac{1}{\alpha(x)}.$$

Hence

$$EH = \int_x \frac{f(x)}{\alpha(x)} dx. \tag{4.2}$$

This is the expression we want to minimise, but there is a normalisation constraint on $\alpha(x)$ which must be met. Observe

$$\alpha(x) = \int_{y=x-\tau/2}^{x+\tau/2} g(y)dy,$$

so

$$\begin{aligned}\int_{x=-\infty}^{\infty} \alpha(x)dx &= \int_{x=-\infty}^{\infty} \int_{y=x-\tau/2}^{x+\tau/2} g(y)dy\, dx \\ &= \int_{x=y-\tau/2}^{y+\tau/2} \int_{y=-\infty}^{\infty} g(y)dy\, dx \\ &= \int_{x=y-\tau/2}^{y+\tau/2} dx \\ &= \tau. \end{aligned} \tag{4.3}$$

It is straightforward now to apply the calculus of variations to minimise equation 4.2 subject to the constraint $\int_x \alpha(x)\, dx = \tau$. The Lagrangian is

$$\frac{f(x)}{\alpha(x)} + \lambda\alpha(x)$$

(where λ is a Lagrange multiplier) and differentiating with respect to α leads to the following Euler-Lagrange equation:

$$0 = -\frac{f(x)}{\alpha(x)^2} + \lambda.$$

That is, we should choose $\alpha \propto \sqrt{f}$, with normalisation determined by the constraint (equation 4.3).

The final step is to express α in terms of g. If g is smooth,[4] we can do this using the Taylor series:

$$\alpha(x) = \int_{y=x-\tau/2}^{x+\tau/2} (g(x) + g'(x)y + \ldots)dy \tag{4.4}$$

$$= g(x)\tau + O(\tau^2). \tag{4.5}$$

Hence

$$g(x) = \frac{1}{\tau}\alpha(x) + O(\tau).$$

So now the question is answered: for sufficiently small τ, the choice of g which minimises EH is $g \propto \sqrt{f}$, normalised of course so that $\int g = 1$.

[4]The approximation is easy to prove if g is smooth, but this is certainly not necessary. Provided only that g is continuous, it's easy to prove a similar formula whose second-order term is proportional to the maximum variation of g over any interval of length τ (often called g's *modulus of continuity* for τ).

Examples

1. **Uniform distribution.** Take $f(x) = 1$ for $x \in [0,1]$ and 0 otherwise. Then $g(x) = 1$ on $[0,1]$ also, and $EH = 1/\tau$. Slogan: "use a uniform distribution to search for a uniform variate".[5]

2. **Semi-triangular distribution.** Take $f(x) = 2x$ for $x \in [0,1]$ and 0 otherwise. Then $g(x) = \frac{3}{2}x^{1/2}$ on $[0,1]$, and $EH = \frac{8}{9}\frac{1}{\tau}$. Note that if we chose to set $g = f$ we would have $EH = 1/\tau$, about 11% worse than optimal. But there is a more grave problem with setting $g = f$ in this case — it turns out that $\operatorname{var}(H) = \infty$. In fact, if we set $g(x) = (p+1)x^p$ on $[0,1]$ then
$$EH = \begin{cases} \frac{1}{\tau}\frac{2}{2+p+p^2} & \text{if } p < 2, \\ \infty & \text{if } p \geq 2, \end{cases}$$
and
$$\operatorname{var}(H) = \begin{cases} \frac{1}{\tau^2}\frac{2(p^2-2p-2)}{(2-p)^2(1-p)(1+p)^2} & \text{if } p < 1, \\ \infty & \text{if } p \geq 1. \end{cases}$$
These expressions were tested experimentally; the results are shown in Figure 4.10.

3. **Normal distribution.** Take $f(x) = \frac{1}{\sqrt{2\pi}}\exp(-\frac{1}{2}x^2)$. Then $g(x) = \frac{1}{2\sqrt{\pi}}\exp(-\frac{1}{4}x^2)$. But it turns out that with this choice of g, $\operatorname{var}(H) = \infty$, and that for practical purposes it is better to use a normal distribution with a higher standard deviation. Indeed, suppose that $g(x) = \frac{1}{\sqrt{2\pi}\sigma}\exp(-\frac{1}{2}\frac{x^2}{\sigma^2})$. Then it is easy to calculate that
$$EH = \begin{cases} \infty & \text{if } \sigma \leq 1, \\ \frac{1}{\tau}\frac{\sigma^2}{\sqrt{\sigma^2-1}} & \text{if } \sigma > 1, \end{cases}$$
and that
$$E(H^2) = \begin{cases} \infty & \text{if } \sigma \leq \sqrt{2}, \\ \frac{4\pi}{\tau^2}\frac{\sigma^3}{\sqrt{\sigma^2-2}} & \text{if } \sigma > \sqrt{2}. \end{cases}$$
Figure 4.11 shows the results of a numerical simulation to verify these calculations. Observe that choosing $\sigma \approx 1.6$ is a good choice because it makes EH only 2.5% higher than the theoretical minimum, and also leads to an acceptably small variance for H. Slogan: "use a normal distribution to search for a normal distribution, but with standard deviation at least 1.5 times bigger".

[5] Actually, because this f violates the continuity assumption so badly, the value of EH obtained in numerical tests is up to 1% higher than the theoretical one. It's not hard to calculate g from α explicitly in this case, thus obtaining a truly optimal sampling function, but the details are unimportant.

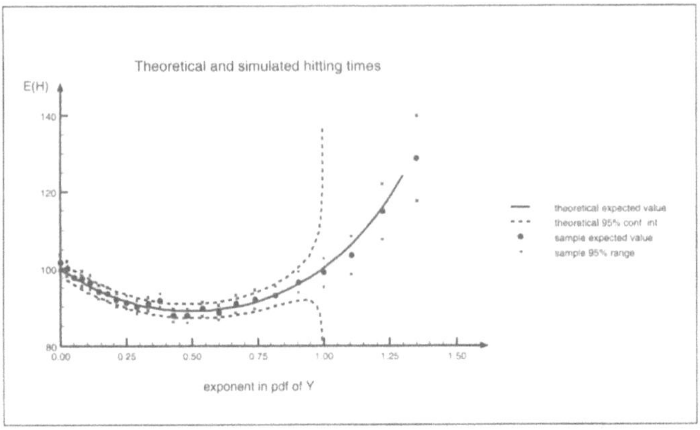

Figure 4.10: ***Hitting times for a semi-triangular distribution.*** *The target's pdf is $f(x) = 2x$ on $[0,1]$, and the sampling pdf is $g(x) = (p+1)x^p$, where p varies along the x-axis. Note that the variance of H becomes infinite at $p = 1$ — precisely the value of p corresponding to the choice $g = f$! Of course for greater values of p, the sample variance remains finite but there is a significant chance of numerical overflows occurring. The minimum value for EH occurs at $p = 0.5$, as predicted in the text. In the region of infinite variance, the sample variance does not blow up as predicted because the runs which caused numerical overflow were disregarded.*

The observations in this section have important implications for the design of searching systems based on random sampling. The system described earlier took account of them only qualitatively, by using priors broader than the true expected prior on the target's configurations. The results of this section would need to be incorporated much more precisely in any attempt to optimise the system's hitting times.

4.6 Tracker initialisation by factored sampling

An important application of object localisation is to initialising contour trackers, or for reinitialising them after they have lost lock on the target. We briefly discuss how this might be done for the two types of tracker: the Kalman filter (Blake et al., 1993; Terzopoulos and Szeliski, 1992), and sampling trackers such as the Condensation algorithm (Chapter 2). The Kalman filter requires a single estimate of the target object's configuration together with the covariance matrix of this estimate; sampling trackers require a fair sample of specified size (typically 50–10,000) from the prior distribution $p(\mathbf{x}|\text{entire image})$.

Both the examples given here are the simplest possible "proof-of-concept" implementations to demonstrate the utility of the contour likelihood ratio in tracker initialisation. Later in the book (sections 4.7 and 7.6.1) more sophisticated systems are described which assimilate these and other ideas to produce

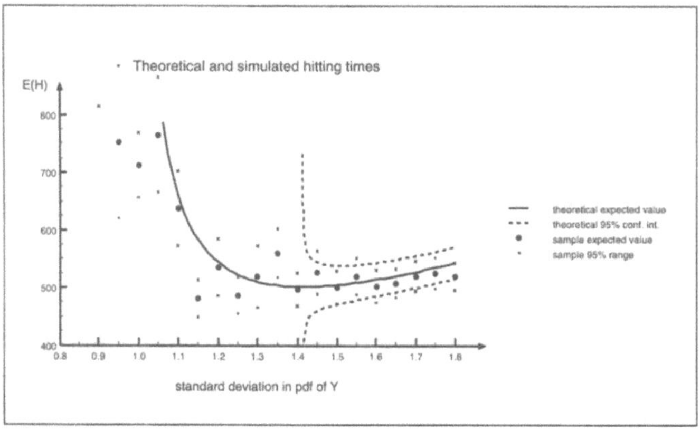

Figure 4.11: ***Hitting times for a normal distribution.*** *Here the target has a standard normal distribution, and the sampling pdf is a zero-mean normal with standard deviation σ, which is the x-axis variable. Note that the "obvious" choice $\sigma = 1$ leads to disaster, since EH is infinite. Moreover, even the "optimal" choice $\sigma = \sqrt{2}$ is bad since* $\text{var}(H)$ *is infinite. A better choice is $\sigma \approx 1.6$. Again, the sample variance does not blow up as predicted because the runs which caused numerical overflow were disregarded.*

robust tracking in challenging conditions.

4.6.1 Kalman filter tracker

An initialiser/re-initialiser for a Kalman filter contour tracker has been implemented, and its typical behaviour is shown in Figure 4.12. The tracker periodically calculates the contour likelihood ratio $\mathcal{R}(\mathbf{x})$ (equation 3.7) of the current configuration. If this is below one, lock is considered to have been lost and the fixed prior is sampled randomly until a configuration with contour likelihood ratio greater than one is found. This configuration is passed back to the Kalman filter, together with a default covariance. Surprisingly good results were obtained, but of course the robustness of the re-initialisation mechanism depends on the properties of $\mathcal{R}$: the property that "$\mathcal{R}(\mathbf{x}) > 1 \implies$ locked on to target" is crucial. By checking the graphs in Figure 3.4, we can see that this property can be relied on for the examples given. However, in the presence of enough clutter, false positives (non-target configurations with $\mathcal{R}(\mathbf{x}) > 1$) will be present and this re-initialisation mechanism has no method of avoiding them.

4.6.2 The Condensation tracker

Initialising a Condensation tracker can be achieved by a straightforward application of the factored sampling object localisation method in Section 4.2. Both algorithms share an emphasis on estimating an entire distribution rather

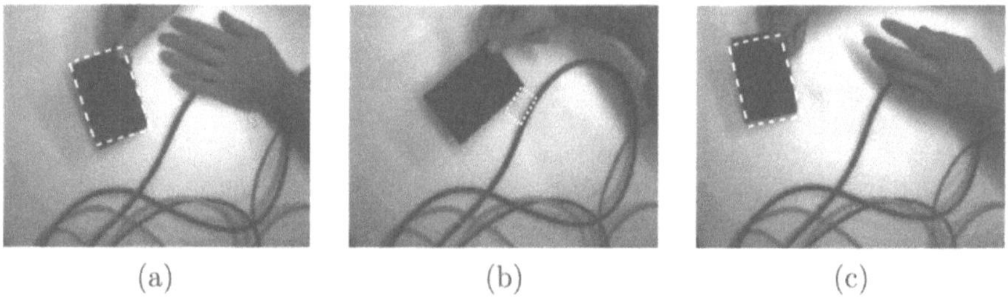

Figure 4.12: ***Kalman filter tracker/reinitialiser.*** *(a) The tracker has been successfully initialised by factored sampling of contours. (b) The tracker loses lock on the target due to distraction by a clutter feature. (c) The system realises lock has been lost and reinitialises using factored sampling, taking less than a second in this case. Parts of the background clutter are also moving (the two hands), but the re-initialisation is unaffected by this.*

than a single mode or mean, and in particular this means both can deal with multi-modal observation densities.

Figure 4.13 gives an example. It shows all particles with contour likelihood ratio greater than 0.1 after sampling from the prior $p(\mathbf{x})$ 10,000 times. Simply passing this set to a Condensation tracker, with weights given by normalised contour likelihoods, serves as a satisfactory initialiser. This approach to initialising Condensation was applied to the simple example shown in Figure 4.14. The mechanism for determining when to reinitialise and with that configuration is the same as for the Kalman filter.

Section 4.7 gives details of a self-initialising head tracking system which extends these ideas to a robust, real-time application.

4.7 Tracking using Condensation and the contour likelihoods

This section describes a robust, real-time system for tracking human head-and-shoulder outlines by fusing colour information with the contour likelihoods of Chapter 3.

4.7.1 The robustified colour contour likelihood

Skin colour is a strong cue which can be used to locate and track humans as they move in a video sequence. A natural extension of the work so far would be to develop a generative model and likelihood function for colour, in the same spirit as the contour likelihoods of Chapter 3. Skin colour has been studied extensively in the vision community, both for its intrinsic properties and for its utility as a tracking cue (e.g. Fleck et al., 1996; Kjeldsen and Kender, 1996; Jones and Rehg, 1998).

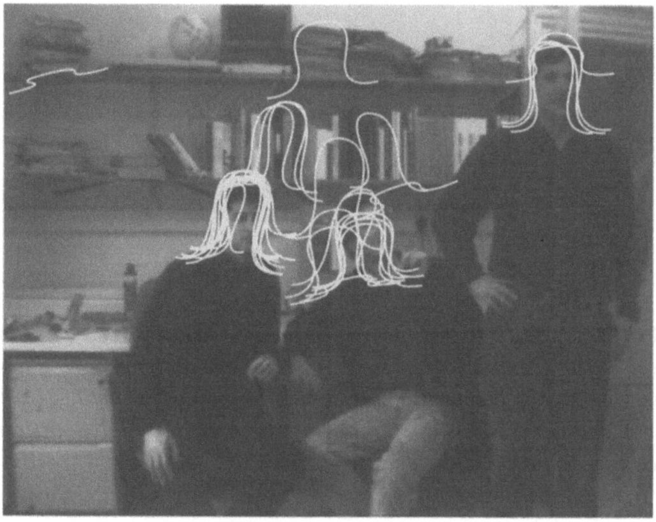

Figure 4.13: ***Initialising a sampling tracker.*** *10,000 particles were taken from $f(w)$, and those with contour likelihood ratio over 10.0 are displayed. (There are about 40 of them). This set, or all 10,000 particles, can be passed directly to the Condensation algorithm (with normalised contour likelihoods as weights) as an initialising set.*

Generative model for RGB values: non-robust version

We first describe a generative model which is extremely simple, and not robust to deviations from the skin-colour model. In this non-robust version, the colour of a single skin-coloured pixel is assumed to be drawn from a truncated Gaussian distribution in RGB-space, learned from hand-segmented training images. The colour of a single pixel in the background is assumed to be drawn from the uniform distribution on the entire space of RGB values. Because pixels are sampled only sparsely, these random colour values are assumed to be independent.

To be more specific, denote the space of possible RGB values by $R = [0, 255]^3$. Suppose $r_1, \ldots r_n$ is a collection of RGB values in R, obtained by sparsely sampling the skin-coloured regions of some training images. In our case, $n \approx 500$. Let $g(r)$ be a Gaussian probability distribution in $\mathbb{R}^3$ with the same mean and covariance as the r_i, but truncated to be non-zero only on R and renormalised. The generative model assumes an isolated skin-coloured pixel will have an RGB value drawn randomly from $g(r)$. On the other hand, let $u(r)$ denote the uniform distribution on R (so $u(r) \equiv 1/255^3$). The generative model assumes an isolated background pixel will have an RGB value drawn randomly from $u(r)$.

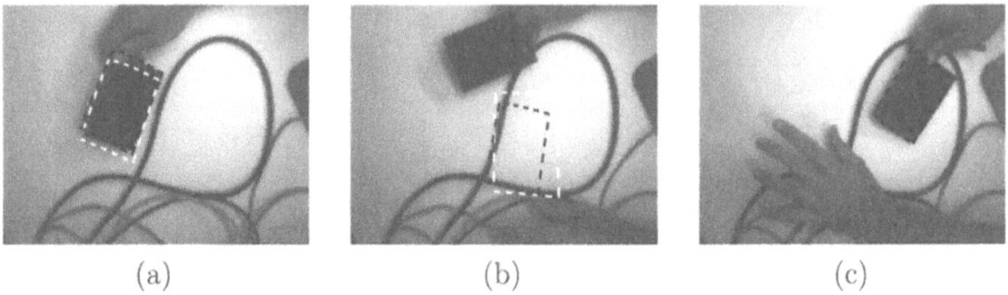

Figure 4.14: ***Reinitialising a Condensation tracker.*** *(a) The black contour is the mean of the Condensation tracker's current distribution. The white contour is the element of the particle set with the highest weight. (b) The tracker has lost lock due to a rapid movement of the target. (c) Re-initialisation has taken place successfully. Once again, parts of the background (both hands) were moving during re-initialisation.*

Generative model for RGB values: robust version

When tracking a head-and-shoulders outline, many pixels on the interior of a correctly positioned contour are not skin-coloured: eyes, mouth, shadows and hair cause deviation from skin colour. One way to deal with this problem would be a much more explicit and detailed model of the object being tracked. Our objective, however, is to achieve real-time performance with a simple model. Thus, a robust version of the generative model above is adopted instead, in which it is assumed that any pixel in the "skin-coloured" foreground region has a fixed probability of resembling a background pixel. That is, a *robustness parameter* $\rho \ll 1$ is fixed, and foreground pixels are assumed to be drawn from the distribution

$$\tilde{g}(r) = \rho u(r) + (1 - \rho) g(r).$$

The results later took $\rho = 0.02$, which was determined to give good results by trial and error.

Likelihood for colour

For this implementation, the pixels used for inference were chosen to be those at the end-point of the interior portion of each measurement line (see Figure 4.4, for example). If the hypothesised position of the contour is correct, then these pixels will be drawn from foreground distribution, and they are sufficiently sparse for the approximation that their RGB values are independent to be acceptable. If there are M measurement lines, then there will be M such RGB values each depending on the hypothesised configuration $\mathbf{x}$ — say $r_1(\mathbf{x}), \ldots r_M(\mathbf{x})$. The *robustified colour likelihood* is defined as

$$\mathcal{P}_{\text{col}}(\mathbf{x}) = \prod_{i=1}^{M} \tilde{g}(r_i(\mathbf{x})). \tag{4.6}$$

Finally, the *robustified colour contour likelihood* is defined as the product of the colour likelihood and the order statistic likelihood:

$$\mathcal{P}_{\text{roc}}(\mathbf{x}) = \mathcal{P}_{\text{col}}(\mathbf{x})\,\mathcal{P}_{\text{ord-stat}}(\mathbf{x}). \tag{4.7}$$

The order statistic likelihood was chosen for this application because it can be calculated rapidly.

Note that a meaningful likelihood ratio combining colour and contour can also be defined. Set the background colour likelihood to be

$$\mathcal{B}_{\text{col}}(\mathbf{x}) = \prod_{i=1}^{M} u(r_i(\mathbf{x})).$$

Then the colour likelihood ratio is $\mathcal{R}_{\text{col}}(\mathbf{x}) = \mathcal{P}_{\text{col}}(\mathbf{x})/\mathcal{B}_{\text{col}}(\mathbf{x})$ and the likelihood ratio is $\mathcal{R}_{\text{roc}}(\mathbf{x}) = \mathcal{R}_{\text{col}}(\mathbf{x})\mathcal{R}_{\text{ord-stat}}(\mathbf{x})$. As in Section 3.1.5, a value of $\mathcal{R}$ greater than one indicates the measurements are more "target-like" than "clutter-like".

4.7.2 Implementation of a head tracker

Initialisation

Automatic initialisation of the head tracker is achieved by repeatedly sampling from the prior of Section 4.4 until a configuration with $\mathcal{R}_{\text{roc}}(\mathbf{x}) > 1$ is found. As soon as such a configuration is found, it is assigned a weight of one and the Condensation algorithm proceeds exactly as described in Chapter 2, using the robustified colour contour likelihood as the observation density. To increase the density of samples, the prior was actually restricted to a preset region about one-quarter the size of the video image. This region was centred on the door of an office scene from which people were expected to enter. Experiment showed that, without this restriction, entries of the target would occasionally be missed.

Tracking

The Condensation algorithm was used with dynamics set by the learning methods of Blake and Isard (1998). The head-and-shoulders template was permitted to deform in an eight-dimensional shape space obtained by principal components analysis. On an SGI O2, with a single R10000 processor, we found that Condensation would run at 50 Hz with 75 particles. For many tracking applications this would be a dangerously low number of particles. However, the additional robustness of the robustified colour contour likelihood, which incorporates colour information, means that the tracker generally does not lose lock even at challenging points in a video sequence. Instead, the tracker tends to sacrifice geometrical accuracy (that is, the precise alignment of the contour with the head and shoulders) while preserving the correct approximate position of the contour since the colour of the sampled interior pixels must follow the generative model. Nevertheless, tracking survives through major deviations from the skin colour model (such as the subject turning through 180°) since the edge information has a dominant effect in such situations.

Results

By combining colour and edge measurements in a simple yet rigorous fashion, this head tracking implementation can correctly track challenging sequences in real time. Figure 4.15 gives an example. It shows stills from a 60-second sequence shot with a hand-held camera. The camera undergoes unknown and unpredictable motion, and the subject moves about the room freely, including turning away from the camera, sitting and rapid transverse motion. No post-processing is performed: the figure shows the output produced on the computer screen in real time as the sequence is shot. As the stills show, lock is maintained on the target throughout the sequence. At times of rapid motion, the accuracy is reduced but the colour component of the model enables the gross motion of the subject to be estimated correctly. The tracked video sequence itself can be viewed at the book web site (see page 3).

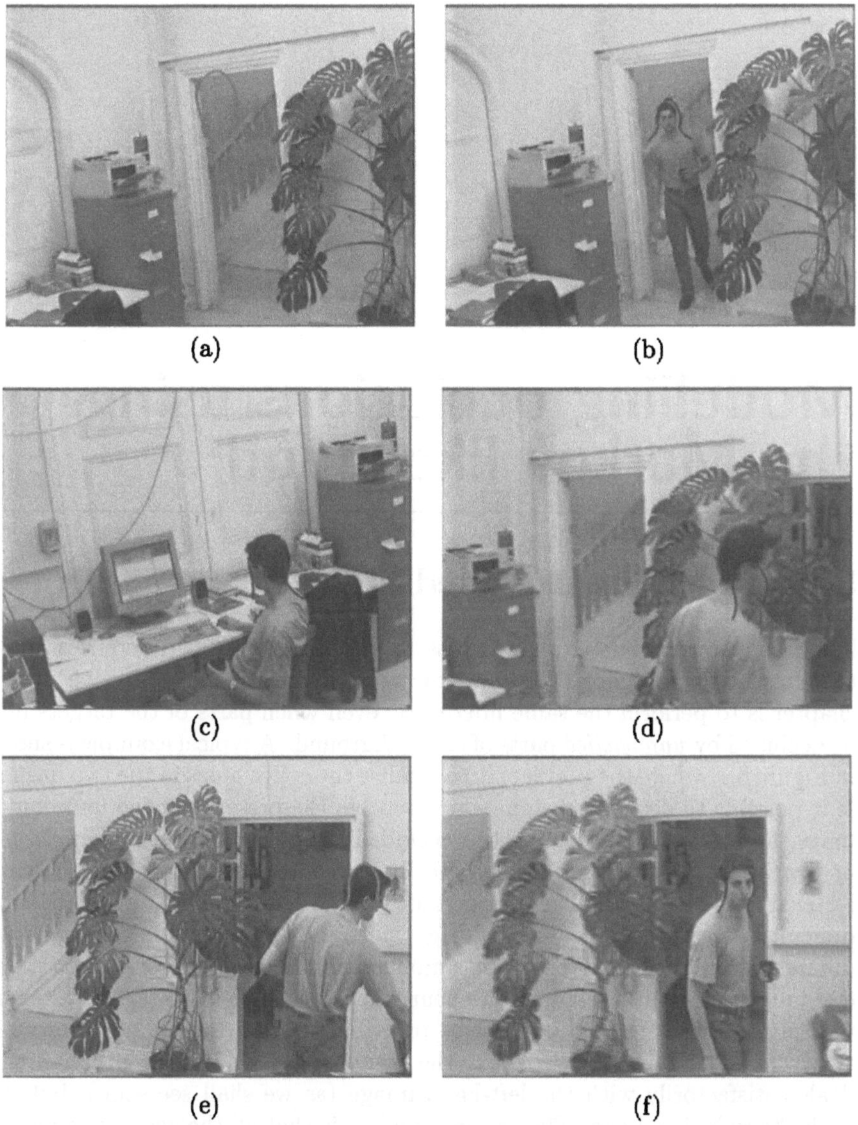

Figure 4.15: ***Head tracking results.*** *Typical frames from a 60-second real-time sequence with a hand-held moving camera and freely moving subject. The tracked sequence itself is available from the book web site (see page 3). A light-coloured contour indicates $\mathcal{R}_{\rm roc}(\overline{\mathbf{x}}) < 1$; dark-coloured implies $\mathcal{R}_{\rm roc}(\overline{\mathbf{x}}) \geq 1$. (a) Initialisation mode. The light-coloured contour is just visible in the door frame. (b) Immediately after initialising correctly. The dark contour is locked onto the subject. (c) Continued correct tracking, despite major deviation from skin-colour model since most of the interior of the contour consists of the subject's dark hair. (d) During rapid motion, tracking continues but is much less accurate; the skin-coloured region receives high likelihoods and thus aids robustness. (e) Again the skin model fails but edge measurements enable tracking to continue. (f) Accurate tracking resumes.*

5

Modelling occlusions using the Markov likelihood

5.1 Detecting occluded objects

Previous chapters have explained how to perform statistical inferences about the presence or absence of target objects in a scene. The objective of this chapter is to perform the same inferences, even when parts of the targets may be occluded by *unmodelled* parts of the background. A typical example is shown in Figure 5.1, where the problem is to localise the coffee mugs in the two images. Is it possible to design a system which reports the presence of the unoccluded mugs, and in addition detects the occluded mug *with an appropriate degree of confidence*? Note that a heuristically based recognition system (relying, for example, on the number of a certain type of feature matches) might have difficulty even with the left-hand image, since the two targets might have very different scores. This problem is amplified in the right-hand image, where one mug is partially occluded: the heuristic scores of the two targets are very unlikely to reflect the actual relative probabilities that targets are present in those two configurations. In fact, the contour likelihood ratio of Chapter 3 deals satisfactorily with the left-hand image (as we shall see soon), but not with the right-hand one. The crucial problem is that all the generative models of Chapter 3 assume outputs on distinct measurement lines are *independent*. When many consecutive measurement lines are occluded by an unmodelled part of the "background",[1] this independence assumption is violated badly, and unrealistic inferences are the result.

An essential component of systems which can output relative probabilities is a stochastic generative model of how the measured image features are generated. The ideas of other authors (e.g. Grimson et al., 1992; Leung et al., 1995; Lowe, 1992; Shimshoni and Ponce, 1995) were surveyed in Chapter 4,

[1] Here the word "background" is used in the sense of background clutter, which is taken to be everything other than the target objects. Hence it makes sense to talk of the background occluding the targets.

(a) cups with no occlusion (b) one cup partially occluded

Figure 5.1: ***Can a localisation system produce meaningful results from scenes like these?*** *A heuristically based system may or may not be able to detect both mugs in both images after appropriate tuning. However, the real challenge is to report a realistic probability that the partially occluded mug in (b) is indeed a target. Heuristic scoring functions are of little use in answering this challenge.*

but none of these address the specific problem of interest in this chapter, which is to obtain realistic inferences despite occlusion. Amir and Lindenbaum (1996) proposed a powerful method for assessing partially occluded targets, which used graph partitioning and "grouping cues" to draw inferences on whether missing edge information is due to occlusion. Although their model was designed entirely in terms of elementary probabilities, the output was chiefly useful for identifying a single best hypothesis. Indeed, the likelihoods for plausible configurations tended to differ by many orders of magnitude, possibly due to the assumption of independence between grouping cue measures.[2] Another effective approach was suggested by Rothwell (1996). This used image topology, and T-junctions in particular, to assess whether missing boundaries were genuine occlusions. It is not clear to what extent the "verification scores" of Rothwell's work can be used for statistical inferences, but, in any case, the methodology presented here uses measurements based on a different set of image features. Hence the outputs of each system could in principle be fused to achieve even better performance. The "Markov likelihood" described in this chapter uses a cyclic Ising model for occlusions on target boundaries. A similar approach was developed independently by Rue and Husby (1997), who used the cyclic Ising model in the context of Markov chain Monte Carlo simulations on medical images.

[2]If this is indeed the reason for this effect, then it makes an interesting comparison with the likelihoods discussed in this chapter, since the Markov likelihood is designed to eliminate an independence assumption from the contour likelihood ratio framework.

(a) output using an independence generative model on unoccluded mugs

(b) output using an independence generative model with one mug partially occluded

Figure 5.2: ***Relative values of the likelihood when using independence generative models are not realistic for partial occlusions.*** *Posteriors are estimated by the factored sampling method of Chapter 4. The independence generative model performs adequately in (a), producing peaks in the posterior with similar magnitudes at each target. However, the results in (b), in which one target is partially occluded, are not at all realistic. The displayed intensity of each contour in these figures is proportional to the log of its likelihood. The right-hand peak in (b) is actually* 10^{-8} *times weaker than the left-hand peak and would be invisible on a linear scale. Note carefully the sense in which the independence generative model has failed. Both mugs are successfully detected, as there are strong peaks in the posterior at each target. However, the magnitudes of these peaks are not realistic when interpreted as relative probabilities.*

5.2 The problem with the independence assumption

A crucial assumption in all the generative models of Chapter 3 was Assumption 6 (page 41): that feature outputs on distinct measurement lines are statistically independent. For convenience, let us call all of those models *independence generative models.* This chapter will introduce a new class of models called *Markov generative models*, by replacing Assumption 6 by something more appropriate that involves a Markov random field (MRF).

To understand why the independence assumption can lead to unrealistic results, suppose a small but significant portion of the target outline is occluded — up to seven or eight measurement lines, for instance. Then the likelihood $\mathcal{P}(\mathbf{Z}|\mathbf{x})$ of the measurements $\mathbf{Z}$ is reduced dramatically, since according to our model of the feature formation process, seven or eight unlikely events have occurred independently. An example is shown in Figure 5.2. In fact, only one unlikely event has occurred — a single interval of contour was occluded — but the likelihood of the independence generative model does not reflect this.

5.3 The Markov generative model

To solve the problem seen in Figure 5.2, we need a model reflecting the fact that occlusion events on nearby measurement lines are not independent. More specifically, the incorporation of such a model in a Bayesian framework will require a prior expressing the type and amount of occlusion expected. The approach taken here is to express the prior as an MRF, regarding the measurement lines round the contour[3] as the sites of the MRF. The possible states of each site are "visible" and "occluded". Formally, suppose there are M measurement lines and denote the state of site i by s_i. We adopt the convention that $s_i = 1$ if site i is occluded, and $s_i = 0$ if site i is visible. An entire state vector $(s_1, \dots s_M)$ will normally be denoted just by $\mathbf{s}$, and the set of all possible values of $\mathbf{s}$ is written $\mathcal{S}$; note that $\mathcal{S}$ has 2^M elements. The prior on $\mathcal{S}$ is denoted Θ, and the next section describes how the values of $\Theta(\mathbf{s})$ were determined in our examples.

Meanwhile, we continue the derivation of the new Markov likelihood. Recall some notation from Chapter 3: $\tilde{p}_c(n; \mathbf{z}|\boldsymbol{\nu})$ is the pdf for a measurement line with intersection number c and intersection innovation $\boldsymbol{\nu}$. The basic idea behind the new generative model is that intersections (i.e. target boundaries) on occluded measurement lines are invisible; these measurement lines should have their intersection number set to zero. Formally, suppose the target object is in configuration $\mathbf{x}$. Then the Markov generative model for the formation of edge features on the whole set of measurement lines $\mathcal{L}_1, \dots \mathcal{L}_M$ is as follows:

1. A (generally small, possibly empty) subset of the measurement lines is selected as the occluded measurement lines, according to the prior Θ described in the next section. In other words, we draw a value of the occlusion state vector $\mathbf{s}$ from the prior $\Theta(\mathbf{s})$.

2. On each unoccluded measurement line $\mathcal{L}_i$, the feature generation process proceeds independently with the pdf $\tilde{p}_{c_{\mathbf{x}}(i)}(n; \mathbf{z}|\boldsymbol{\nu}_{\mathbf{x}}(i))$ described in Chapter 3.

3. On each occluded measurement line $\mathcal{L}_i$, the feature generation process proceeds independently[4] with the pdf $\tilde{p}_0(n; \mathbf{z}|\boldsymbol{\nu}_{\mathbf{x}}(i))$ described in Chapter 3. That is, the intersection number c is set to zero on the occluded measurement lines.

Let $\mathcal{P}_{\mathrm{M}}$ be the new likelihood, called the *Markov likelihood*, arising from this generative model. The formula for $\mathcal{P}_{\mathrm{M}}$ is obtained in just the same way as for the generative likelihood $\mathcal{P}_{\mathbf{gen}}$ in Proposition 3.3, equation 3.3. Fix a value of the occlusion state vector $\mathbf{s}$. Then, using the same notation as equation 3.3 we

[3]The model described in this chapter is appropriate for that attached measurement line regime (Section 3.2.3). The same ideas apply to the fixed regime but require some non-trivial modifications to work correctly.

[4]These processes are also assumed to be independent of the processes on unoccluded measurement lines.

have

$$\mathcal{P}_{\mathrm{M}}(\mathbf{Z}|\mathbf{z},\mathbf{s}) = \left(\prod_{i \text{ s.t. } s_i=0} \tilde{p}_0(\mathbf{z}^{(i)}|\boldsymbol{\nu}_{\mathbf{x}}(i)) \right) \left(\prod_{i \text{ s.t. } s_i=1} \tilde{p}_{c_{\mathbf{x}}(i)}(\mathbf{z}^{(i)},\boldsymbol{\nu}_{\mathbf{x}}(i)) \right). \quad (5.1)$$

Of course, to obtain an expression which can be used in calculating a new likelihood, we must sum over all values of $\mathbf{s}$, weighting by the prior probabilities $\Theta(\mathbf{s})$. This gives

$$\mathcal{P}_{\mathrm{M}}(\mathbf{Z}|\mathbf{x}) = \sum_{\mathbf{s}\in\mathcal{S}} \mathcal{P}_{\mathrm{M}}(\mathbf{Z}|\mathbf{x},\mathbf{s})\Theta(\mathbf{s}). \quad (5.2)$$

There is a crucial difficulty in calculating the Markov likelihood: the sum in equation 5.2 contains 2^M elements (recall M is the number of measurement lines) — far too many to be enumerated explicitly for typical values of M which are 10–100. Hence we must resort to a simulation technique, or equivalently, a Monte Carlo integration. The Monte Carlo method (Mackay, 1999) would be to draw samples $\mathbf{s}^{(1)}, \ldots \mathbf{s}^{(K)}$ from Θ, and estimate equation 5.2 as

$$\frac{1}{K}\sum_{k=1}^{K} \mathcal{P}_{\mathrm{M}}(\mathbf{Z}|\mathbf{x},\mathbf{s}^{(k)}). \quad (5.3)$$

In fact, the convergence of this method is generally not rapid enough for practical estimation of the Markov likelihood. The results in this chapter were instead calculated using an importance sampling technique described in Section 5.7. Section 5.8 outlines the method of using MCMC simulation to obtain the random samples $s^{(k)}$.

5.4 Prior for occlusions

As explained in the previous section, the prior that models the types of occlusion expected will be expressed as an MRF whose sites are the measurement lines and whose state at each site is either "visible" or "occluded". Recall the notation for this: $\mathbf{s} = (s_1, \ldots s_M)$ is a state vector of the MRF, with $s_i = 1$ if the nth measurement line is occluded and 0 otherwise. The set of all possible values of $\mathbf{s}$ is $\mathcal{S}$.

Our objective in this section is to define a prior Θ on $\mathcal{S}$. As explained in texts on MRFs (e.g. Winkler, 1995), there is a one-to-one correspondence between priors Θ and energy functions H. This correspondence is given by

$$\Theta(\mathbf{s}) = Z^{-1}\exp(-H(\mathbf{s})), \;\; Z = \sum_{\mathbf{s}'\in\mathcal{S}} \exp(-H(\mathbf{s}')). \quad (5.4)$$

Configurations with higher energy have lower prior probability, as in thermodynamics. The method for designing such a prior has two steps. First, fix the functional form of a suitable energy function by incorporating intuitive notions of its desirable properties, and second, select or learn any parameters to achieve good behaviour for a given target object. The intuitive ideas to be incorporated by the first step are:

(a) extensive occlusion is relatively unlikely;

(b) the occlusion is more likely to occur in a small number of contiguous intervals than in many separated intervals.

These are expressed by an energy function H of the form

$$H = \alpha \sum_{i=1}^{M} s_i - \beta \sum_{i=1}^{M} s_i s_{i+1}, \tag{5.5}$$

where α and β are positive real parameters to be determined later.[5] The first term in this expression penalises every occluded site, thus incorporating the intuitive idea (a) above. The second term encourages occlusion at adjacent sites, incorporating intuitive idea (b). It can be made even more explicit that idea (b) really has been captured here. Let $O = \sum s_i$ be the number of occluded sites and let I be the number of contiguous occluded intervals. Then penalising each of these quantities in the energy function would suggest adopting $H = \alpha' O + \beta' I$, for some α', β'. But observe that $I = O - P$, where $P = \sum s_i s_{i+1}$ is the number of adjacent pairs of occluded sites. Hence $H = (\alpha' + \beta')O - \beta' P$, exactly as in equation 5.5 if we take $\alpha = \alpha' + \beta'$ and $\beta = \beta'$.

The choice of precisely how to incorporate the two intuitive ideas is of course rather arbitrary. The above choice was guided by the desirability of simplicity. Note that the first term of equation 5.5 is the sum of single-site potentials, and the second term is the sum of pair potentials. This can be immediately recognised as the energy for an Ising model, and the graph of its neighbourhood system is the "necklace" shown in Figure 5.3 — this is sometimes called a cyclic MRF (Kent et al., 1996). Quantities of interest can now be calculated easily. For example, it turns out the probability of occlusion given that the two neighbours of a site are visible is given by

$$\text{Prob}(s_i = 1 \mid s_{i-1} = s_{i+1} = 0) = (1 + \exp(\alpha))^{-1}, \tag{5.6}$$

and the probability of an occluded site between two other occluded sites is

$$\text{Prob}(s_i = 1 \mid s_{i-1} = s_{i+1} = 1) = (1 + \exp(\alpha - 2\beta))^{-1}. \tag{5.7}$$

In the examples shown here, the parameters α, β were chosen so that the expected number of occluded sites is 5% of the total number of sites, and the expected number of contiguous intervals of occluded sites is 0.7; this corresponds to $\alpha = 5.33$ and $\beta = 5.0$. Alternatively, the values of α, β could, in principle, be learned from training data. A random sample from a simulation of this prior is shown in Figure 5.3.

It is worth addressing one further question here: is it possible to specify α and β as a function of M, the total number of sites, in such a way that some desirable statistical properties are constant? If so, then this more general approach would be preferable to finding suitable α, β numerically for each new

[5] We adopt the convention $s_{N+1} = s_1$.

Figure 5.3: ***The prior for occlusions.*** *Each small dot is a site in the MRF, and the local characteristics of each site depend only on its two neighbours. In the application described here, each "site" is actually a measurement line on the contour (see Figure 1.2, for example), and the possible states of each site are "occluded" or "visible". Nine samples from the MRF described in the text are shown here; grey circles are visible and black circles are occluded. This MRF has 44 sites, the same as the number of measurement lines in the coffee mug template (see Figure 5.6).*

class of target. Unfortunately, this turns out to be a rather difficult problem. Statistical physicists are interested in the same question, and attempts to answer it have led to the theory of renormalisation group transformations (Chandler, 1987). Vision researchers have also used renormalisation theory, though in a completely different context to the present problem (Gidas, 1989; Perez and Heitz, 1996). Some unpublished work[6] suggests that numerical schemes for altering the parameters α, β may be of some use, but this is unnecessary for the examples addressed here, since M is fixed in advance for any given localisation problem.[7]

[6]Geoff Nicholls, personal communication, 1997.

[7]I am indebted to Geoff Nicholls who pointed out the connection to renormalisation group theory and suggested relevant literature.

(a) output using Markov likelihood with unoccluded mugs

(b) output using Markov likelihood with one mug partially occluded

Figure 5.4: ***The Markov likelihood produces more realistic likelihoods.*** *When both mugs are unoccluded (a), the peak value of the likelihood at the left-hand mug is about three times that at the right-hand mug. When the right-hand mug is occluded (b), the peaks differ by a factor of about four. Both these results are more realistic than those calculated by the independence likelihood (Figure 5.2), but the improvement is particularly marked in the case of partial occlusion.*

5.5 Realistic assessment of multiple targets

Section 5.5.1 gives a broad outline and explanation of the result shown in Figure 5.4. Section 5.5.2 gives precise details of how the experiment was carried out.

5.5.1 Explanation of results

Recall our objective in introducing the Markov likelihood: to obtain roughly equal peaks in the likelihood when evaluated at true target configurations, regardless of whether a small portion of the outline is occluded.

To test this we applied the method to two nearly identical scenes, Figure 5.1(a) and (b). The first scene shows two coffee mugs with their handles visible; the second shows the same scene, but this time the handle of the right-hand mug is occluded by a hand which is about to pick up the mug. Figure 5.2(a) and (b) shows the posterior distribution as calculated by the independence likelihood. (Section 5.5.2 describes in detail exactly what these figures represent and how they were created.) When neither mug is occluded, the two peaks in the distribution differ by a factor of about 22 — not a particularly realistic result, but at least the peaks have similar orders of magnitude. However, when the right-hand mug has its handle occluded (Figure 5.2(b)), the independence likelihood evaluates the corresponding peak in the posterior as being approximately 10^{-8} times smaller in magnitude than the peak for the left-hand mug! This is essentially because eight measurement lines were occluded, and since the non-detection probability q_{01} was set to 0.1, the inde-

	ratio of independence likelihoods	ratio of Markov likelihoods
both mugs visible	22	0.31
one mug partially occluded	5.8×10^{-8}	0.24

Figure 5.5: ***Relative heights of peaks in posterior distributions shown in Figures 5.2 and 5.4.*** *The figures shown are the peak value of the likelihood at the right-hand mug divided by the peak value at the left-hand mug. The ratios of peak values for the Markov likelihood are much more realistic than those given by the independence likelihood.*

pendence likelihood considers that eight independent events of probability 0.1 have occurred.

Next the Markov likelihood was applied to the same two scenes (Figure 5.4). As before, when neither mug is occluded the two peaks in the posterior are of similar magnitude. However, Figure 5.4(b) shows that the two peaks still have a similar magnitude even when the handle of the right-hand mug is occluded. This is because, according to the MRF prior we used on the occlusion status vectors of the measurement lines, the event that 8 consecutive measurement lines are occluded is not considered particularly unlikely.

More precise details of the relative values of the peaks in the likelihoods are given in Figure 5.5. Note that these numbers can actually be used for the type of statistical inferences discussed in Section 4.3. For example, suppose for simplicity we have prior knowledge that precisely one mug is present in the scene. Let $\mathbf{x}_1$ be the configuration at the left-hand peak in the posterior and $\mathbf{x}_2$ the configuration at the right-hand peak. Then

$$\begin{aligned}
\text{Prob}(\mathbf{x} = \mathbf{x}_1 | \mathbf{x} = \text{either } \mathbf{x}_1 \text{ or } \mathbf{x}_2) &= \frac{\mathcal{P}(\mathbf{Z}|\mathbf{x}_1)}{\mathcal{P}(\mathbf{Z}|\mathbf{x}_1) + \mathcal{P}(\mathbf{Z}|\mathbf{x}_2)} \\
&= \frac{\mathcal{P}(\mathbf{Z}|\mathbf{x}_1)/\mathcal{P}(\mathbf{Z}|\mathbf{x}_2)}{\mathcal{P}(\mathbf{Z}|\mathbf{x}_1)/\mathcal{P}(\mathbf{Z}|\mathbf{x}_2) + 1} \\
&= \frac{1/0.24}{1/0.24 + 1} \\
&= 0.81.
\end{aligned}$$

Similar calculations can be made with more realistic priors on the number of mugs present in the scene (see also Section 4.3).

5.5.2 Experimental details

The coffee mug template was taken from the scene in Figure 5.6; note that this is a different mug and background to those used for the experiment. The prior used for mug configurations had the following properties: uniform distribution over the two Euclidean translation parameters; rotation in the plane by an angle whose mean is zero and standard deviation 3°; scaling in the x-direction by

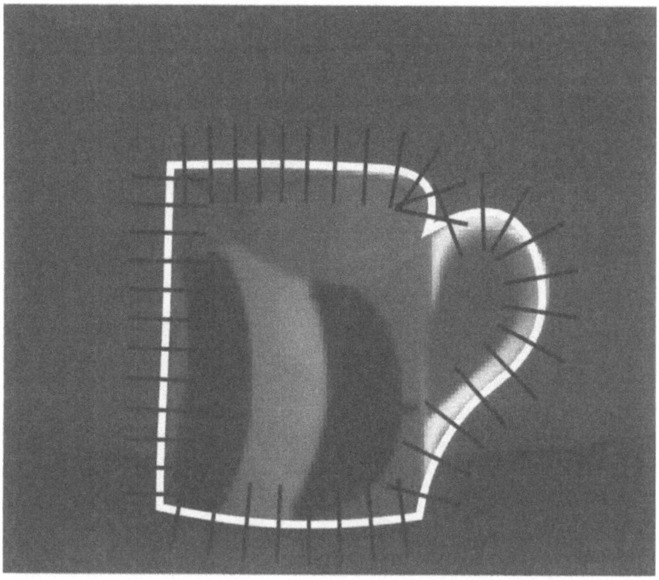

Figure 5.6: ***Template for coffee mug experiment.*** *There are 44 measurement lines, which means 44 sites in the cyclic MRF used to specify the prior on which parts of the contour will be occluded.*

a normally distributed factor whose mean is one and standard deviation 0.1; scaling in the y-direction by a normally distributed factor whose mean is one and standard deviation 0.05. Likelihoods were based on the "generic" generative model of Section 3.1.1, with $q_{01} = 0.1$ and with background probabilities $b(n)$ set to 0.05 for $n < 20$ and zero otherwise. For each scene, the same 10,000 samples were drawn from this prior, and the independence likelihood evaluated for each one. The 100 configurations with the highest independence likelihoods were recorded for further investigation, and the remainder discarded. This approach was taken mainly for a practical reason: it takes several seconds to estimate the Markov likelihood, whereas the independence likelihood has a closed-form formula which can be calculated in milliseconds.

The Markov likelihoods of the selected 100 configurations were then estimated by the importance sampling method described in Section 5.7. In the plots of the posterior distributions shown in figures 5.1, 5.2, and 5.4, the total mass of each contour (intensity $\times$ width) is proportional to the log of the contour's likelihood.

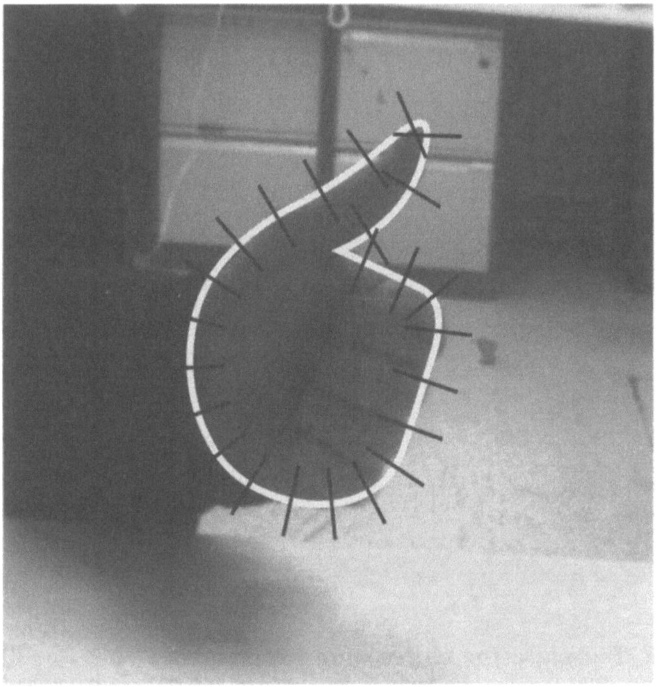

Figure 5.7: ***Template for "thumbs-up" gesture.*** *Note that the signaller is wearing long-sleeved clothing so there are detectable edges on the wrist area.*

5.6 Improved discrimination with a single target

The main motivation for introducing the Markov likelihood was to obtain more realistic relative values in the peaks of the posterior distribution when two or more targets are present. However, experiments showed that in some cases the performance of the method was significantly improved even when only a single target was present.

Figure 5.8 shows an example of this behaviour, where the factored sampling object localisation method is being used to search for a "thumbs-up" signal in a static grey-scale image. The template thumb-signal is shown in Figure 5.7; note that there is good contrast on the wrist because the signaller is wearing long-sleeved clothing. The experiment involved searching for a thumb-signal when the signaller might be wearing short sleeves, in which case no edges would be detected on the wrist. This absence of edge features is actually a generalised form of occlusion.

On around 90% of images tested, both the independence likelihood and the Markov likelihood correctly identified the single target as the strongest peak in the posterior. This type of behaviour is shown in Figure 5.8(a)–(d).

However, occasionally some background clutter is scored higher than the true configuration by the independence likelihood, as in Figure 5.8(e). Of course, the reason the independence likelihood fails is that it finds no edges on the wrist area and consequently gives the true configuration a low likelihood. By evaluating the contours using the Markov likelihood instead, this situation can be rectified: in Figure 5.8(f), for example, the peak at the correct configuration is five times stronger than the one at the spurious hypothesis.

The reason for the improved discrimination is as follows. Recall that the independence likelihood is calculated by multiplying together the likelihoods $\tilde{p}_c(\mathbf{z}^{(i)}|\boldsymbol{\nu}_{\mathbf{x}}(i))$ of the measurements on each individual measurement line. Therefore, the positioning of "unlikely" measurement lines (i.e. those which do not resemble the target) is irrelevant. For instance, a configuration with three very unlikely measurement lines will score the same regardless of whether these three lines are adjacent or separated from each other by intervening sites. The Markov likelihood, on the other hand, takes precisely this type of positioning into account: consecutive (or even nearby) unlikely measurement lines do not incur as great a penalty as separated ones. Consider Figure 5.8(e) and (f) as an example: the two competing configurations have a similar number of poorly scoring measurement lines, but on the true configuration these are all on the wrist, on consecutive lines. The Markov likelihood takes this into account and gives this configuration a higher likelihood.

5.7 Faster convergence using importance sampling

Because it is a product of M factors, the likelihood ratio $\mathcal{P}_{\mathrm{M}}(\mathbf{Z}|\mathbf{x},\mathbf{s})$ defined by equation 5.1 is very sharply peaked when regarded as a function of $\mathbf{s}$. Hence the Monte Carlo estimate (equation 5.3) is dominated by the few samples near a strong peak in $\mathcal{P}_{\mathrm{M}}(\mathbf{Z}|\mathbf{x},\mathbf{s})$, and the majority of samples contribute virtually nothing to the estimate. A standard method to reduce the variance of Monte Carlo estimates is called *importance sampling*[8] (Ripley, 1987). The basic idea is to spend more time sampling near the peaks of $\mathcal{P}_{\mathrm{M}}(\mathbf{Z}|\mathbf{x},\mathbf{s})$, and compensate for this by weighting the calculation appropriately. More specifically, suppose the samples $\mathbf{s}^{(1)},\ldots\mathbf{s}^{(K)}$ are drawn from an importance distribution $\Theta'(\mathbf{s})$ on $\mathcal{S}$. Then as $K\to\infty$ the quantity given by equation 5.2 can be estimated by

$$\frac{1}{K}\sum_{k=1}^{K}\left(\mathcal{P}_{\mathrm{M}}(\mathbf{Z}|\mathbf{x},\mathbf{s}^{(k)}))\times\frac{\Theta(\mathbf{s}^{(k)})}{\Theta'(\mathbf{s}^{(k)})}\right). \tag{5.8}$$

This is true for essentially any choice of $\Theta'(\mathbf{s})$, but of course the idea is to obtain faster convergence, and to this end Θ' should be chosen so that a higher proportion of the samples contribute significantly to the estimate.

[8]In the specific case of statistical mechanical Gibbs samplers, importance sampling is sometimes called non-Boltzmann sampling. An accessible survey of the techniques involved is given in Chandler (1987).

(a) independence likelihood

(b) Markov likelihood

(c) independence likelihood

(d) Markov likelihood

(e) independence likelihood

(f) Markov likelihood

Figure 5.8: ***Improved discrimination with a single target.*** *In the first two examples (a and b, c and d), the independence likelihood and Markov likelihood have similar performance, but the Markov likelihood is significantly better than the independence likelihood at assessing partial matches, as in (e) and (f). Note the reason for the independence likelihood's failure here: the thumbs-up template had strong edges on the wrist, which are not present in (e). However, the Markov likelihood recognises that this generalised "occlusion" of three consecutive measurement lines on the wrist is not particularly unlikely, and therefore does not penalise the true configuration unduly.*

In the particular case of the Markov likelihood, our choice of Θ' is guided by the following observation: the likelihood ratios $\tilde{p}_1(\mathbf{z}^{(i)}|\boldsymbol{\nu}_{\mathbf{x}}(i))/\tilde{p}_0(\mathbf{z}^{(i)})$ of the individual measurement lines give very useful guidance on likely sites of occlusion. Consider a site i for which the value of $\tilde{p}_1/\tilde{p}_0$ happens to be very low. Then it is very plausible that the target boundary was occluded at this site. Hence the importance function Θ' should be biased towards the possibility that site i is occluded. We will say in this case that site i is "encouraged" to be occluded. Of course, this choice can be justified by a purely numerical argument, without referring to whether site i is actually occluded or not. Consider equation 5.2, and again suppose that for a specific site i we have $p_1 \ll p_0$. Then terms in the sum of equation 5.2 for $\mathcal{P}_{\mathrm{M}}$ will be non-negligible only for values of $\mathbf{s}$ that specify the ith site is occluded. Hence we are led to the following conclusion: if, for a fixed configuration $\mathbf{x}$, there is a site i for which $p_1 \ll p_0$, then the importance distribution $\Theta'(\mathbf{s})$ should strongly favour values of $\mathbf{s}$ with $s_i = 1$ (i.e. site i is occluded).

First we discuss the situation when only one site is "encouraged" in this way — the importance sampling method can easily be extended to encourage multiple sites but the notation becomes more complicated. Note that a different encouraged site i is selected for each configuration $\mathbf{x}$. In our implementation this was done by selecting the central site of the longest contiguous low-scoring interval of contour. A measurement line was labelled as low-scoring if the likelihood ratio $\tilde{p}_1/\tilde{p}_0$ was less than a parameter λ. In the rare event that no measurement line was low-scoring according to this definition, we selected the one with the lowest value of $\tilde{p}_1/\tilde{p}_0$. For the examples shown here we took $\lambda = 0.2$, which corresponds to an occlusion probability of 5/6.

The importance function for one encouraged site will be denoted Θ' and is a distribution of the Gibbs form (equation 5.4) with energy denoted by H'. To define H', we must fix in advance a parameter $\gamma > 0$, which determines to what extent occlusion will be favoured at an encouraged site — this choice is discussed briefly below. Then, for a given configuration $\mathbf{x}$, we select by the heuristic above a site i whose occlusion will be encouraged and define the energy of the importance function by

$$H'(\mathbf{s}) = H(\mathbf{s}) - \gamma s_i,$$

where $H(\mathbf{s})$ is the energy of the occlusion prior $\Theta(\mathbf{s})$ defined in equation 5.5. Observe that this choice does indeed "encourage" occlusion at the site i: when $s_i = 1$, the value of H' is reduced by γ, resulting in a more probable configuration.

Note that in order to perform importance sampling using equation 5.8, we must have an expression for $\Theta(\mathbf{s})/\Theta'(\mathbf{s})$. In this case, if we let Z' be the partition function — as defined by equation 5.4 — of the Gibbs distribution with energy H', we have

$$\frac{\Theta(\mathbf{s})}{\Theta'(\mathbf{s})} = \frac{\exp(-H)}{Z} \Big/ \frac{\exp(-H')}{Z'}$$
$$= \frac{Z'}{Z}\exp(-\gamma s_i). \tag{5.9}$$

It remains to explain how to calculate Z'/Z. It turns out that for the simple one-dimensional, two-state cyclic MRFs used in this chapter, all the partition functions can be calculated exactly from a recursive formula, but the ratio Z'/Z can also be estimated for much more general MRFs by a Monte Carlo integration. The details are given in Section 5.9. Note that although we preselected a specific site i at which occlusion would be encouraged, the value of Z' does not depend on i. This is because there is an obvious isomorphism between the MRF at which site i is encouraged and the MRF at which site j is encouraged — just rotate the "necklace" (Figure 5.3) by $i - j$ sites.

Now suppose we wished to improve this importance sampling approach by using an importance function which encouraged occlusion at two different sites i and j. The argument works as before: the importance function Θ'' is of Gibbs form with energy $H'' = H + \gamma(s_i + s_j)$, and a formula analogous to equation 5.9 holds:

$$\frac{\Theta(\mathbf{s})}{\Theta''(\mathbf{s})} = \frac{Z''}{Z}\exp(-\gamma(s_i + s_j)).$$

There is no longer an isomorphism between different choices of (i, j) — in fact the partition function of the importance distribution depends on $|i - j|$. However, for a given value of $|i-j|$, the ratio Z''/Z can be calculated by either of the techniques mentioned above, so the method still works provided the $\lceil M/2 \rceil$ values needed are pre-computed. (Recall that M is the number of sites in the MRF, which is the number of measurement lines on the contour.) This approach can be extended to arbitrary numbers of "encouraged" sites, but the amount of pre-calculation necessary increases as $\binom{M}{E}$, where E is the number of encouraged sites.

The method is valid with any fixed value of γ, but once again the idea is to choose a value which causes equation 5.8 to converge quickly. If Monte Carlo integration is being used to estimate the partition function ratios, then an additional requirement is that these should also converge at an acceptable rate. We have not investigated the issue of how to choose an optimal γ, but empirical tests showed that $\gamma \approx \beta$ worked well. All results presented here took $\gamma = 4.5$. Similar comments apply to the parameter λ: the method is valid for any value, and empirical tests showed that $\lambda = 0.2$ is effective in speeding convergence.

The effectiveness of importance sampling with Θ' and Θ'' is shown in Figure 5.9. Standard Monte Carlo sampling, and the importance sampling method were applied to the best 50 configurations of the unoccluded mug experiment. Each point on the graph is the standard error of 20 estimates of the Markov likelihood, and each of these 20 estimates was calculated from equation 5.8

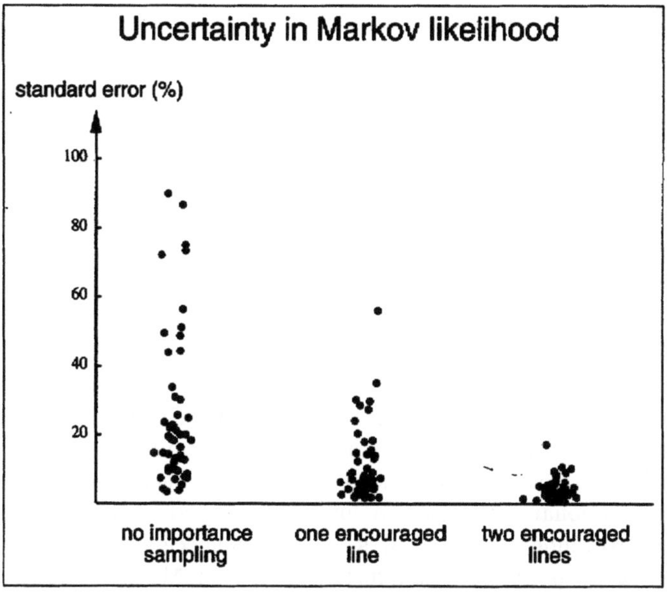

Figure 5.9: ***Importance sampling improves convergence of the Monte Carlo estimates.*** *The Markov likelihood of the best 50 configurations from the unoccluded coffee mug experiment were estimated in three different ways: (1) standard Monte Carlo sampling (i.e. no importance sampling); (2) importance sampling with one "encouraged" measurement line selected for each configuration as described in the text; (3) importance sampling with two encouraged measurement lines. The uncertainty was found empirically by estimating the likelihood value 20 times for each configuration. On the y-axis are shown the standard errors of the 20 estimates, expressed as a percentage of their mean.*

with $K = 20,000$. On average, the standard error of the importance sampling estimate is reduced by 60% if one site at a time is encouraged, and by 85% if pairs of sites are encouraged.

For completeness, here is the heuristic used to select which two measurement lines are encouraged for a given configuration: if there are two or more contiguous low-scoring intervals, select the central site in each of the two longest such intervals. If there is only one contiguous low-scoring interval, select the central site of that interval and the worst-scoring site excluding the interval. If there are no low-scoring lines, select the two with the lowest scores. The definition of "low-scoring" is the same as above.

5.8 Random samples using MCMC

So far we have not said anything about how to draw the random samples $\mathbf{s}_1, \mathbf{s}_2, \ldots$ from $\Theta(\mathbf{s})$ (or Θ', Θ''), which are required for the Monte Carlo estimate of equation 5.3. The only known way of obtaining *independent* samples from an MRF is the algorithm of Propp and Wilson (1996), which uses time reversal and coupling and is rather expensive for general MRFs. Fortunately, however, we don't need independent samples here — equation 5.3 is still valid if the $\mathbf{s}^{(k)}$ are drawn from a Markov chain whose stationary distribution is Θ. MCMC simulation is a well-known method for drawing samples from such a chain. The first example of an MCMC method was by Metropolis et al. (1953), and there are now many good books on the subject, such as Gilks et al. (1996).

The version of MCMC in this chapter uses an arbitrary initial configuration, and visits each site of the MRF in turn, randomly updating it according to the probability distribution for that site, given its neighbours. Equations 5.6 and 5.7 are examples of how to calculate these local update probabilities. After a "burn-in" period of six cycles (i.e. every site has been visited six times) we assume the stationary distribution has been (approximately) achieved, and begin to record random samples $\mathbf{s}_1, \mathbf{s}_2, \ldots$. Each successive sample is obtained from the last by performing six further cycles of MCMC simulation. The assumption is that this is enough to produce a "nearly" independent sample: although the method does not *require* us to use independent samples, it is more efficient to do so, since otherwise many terms in the sum of equation 5.3 would be almost the same.

There is an important open problem associated with the method. This is related to the accuracy and speed of the estimates of the Markov likelihood. The improved performance was gained at the expense of introducing Monte Carlo simulation, which takes seconds rather than milliseconds, and therefore precludes the use of the Markov likelihood in real-time situations. The details given in Section 5.7 show how importance sampling can be used to speed convergence, but even with these improvements it can take over five seconds on a desktop workstation to obtain an estimate with relative error less than 10%. Uncertainties of this magnitude are acceptable for the outputs described here, but better accuracy might be required for more precise statistical inferences. A related problem is therefore the choice of which configurations to analyse with the Markov likelihood, since it is essential that all significant peaks in the posterior are evaluated. For instance, in the experiment with one partially occluded mug, it turns out that the 50 configurations of highest independence likelihood are all clustered about the left-hand unoccluded mug. So evaluating the Markov likelihood at just 50 configurations is not good enough to pick out the peak of nearly equal magnitude at the right-hand mug. A robust solution to this problem might involve clustering the configurations evaluated using the independence likelihood, and ensuring that a representative sample from each cluster was evaluated using the Markov likelihood. Hence one avenue of future work on this topic would be to investigate ways of judiciously choosing when to use MCMC, and for how long.

5.9 Calculating the partition functions

As remarked in Section 5.7, there are two ways to calculate the value Z'/Z required for importance sampling: one is valid for any MRF but provides only an estimate and can be computationally expensive; the other provides an exact solution but can be used only on the simple MRFs described in this chapter.

The first method, which is really just a Monte Carlo integration, relies on the following fact.

Claim Select an "encouraged" site i, and recall that Z' denotes the partition function of the importance function Θ' described in Section 5.7. If $\mathbf{s}^{(1)}, \mathbf{s}^{(2)}, \ldots$ is a sequence of samples from a Markov chain whose stationary distribution is $\Theta(\mathbf{s})$, then

$$\frac{Z}{Z'} = \lim_{K\to\infty} \frac{1}{K} \sum_{k=1}^{K} \exp(-\gamma s_i^{(k)}). \tag{5.10}$$

Proof. By definition, $1 = \sum_{\mathbf{s}\in\mathcal{S}} \Theta(\mathbf{s})$. Regard this as an "integration" of the constant function 1 with respect to the probability measure Θ. We could do this integration using the importance function Θ' instead: select a large K, draw a sample $s_1, \ldots s_K$ from Θ'. Then we expect $1 \approx \frac{1}{K}\sum_{k=1}^{K} \frac{\Theta(\mathbf{s}^{(k)})}{\Theta'(\mathbf{s}^{(k)})}$. Substituting into equation 5.9 gives the result. ∎

Analogous formulas hold if there are two or more encouraged sites. Note that this method of estimating the ratio between two partition functions is similar to enforcing continuity between different regions when carrying out so-called "umbrella sampling" — chapter 6 of Chandler (1987) explains the various possible techniques.

The Monte Carlo estimation approach proved expensive in this application. The desktop workstation used for our experiments took up to 100 seconds to obtain an estimate of Z'/Z whose relative error was less than 5%.

Next we discuss a method for exactly computing the partition functions. The method is an extension of the "transition matrix" method for calculating partition functions of Ising models in statistical mechanics; see chapter 5 of Chandler (1987) for more details on this.

The only trick needed is to maintain four different partition functions of *open-ended* one-dimensional MRFs, defined as follows:

$\Omega_M^{(1,1)} =$ partition function for open chain of M sites starting and ending with 1

$\Omega_M^{(1,0)} =$ partition function for open chain of M sites starting with 1 and ending with 0

$\Omega_M^{(0,1)} =$ partition function for open chain of M sites starting with 0 and ending with 1

$\Omega_M^{(0,0)} =$ partition function for open chain of M sites starting and ending with 1.

For ease of notation, define

$$\Omega_M = \begin{pmatrix} \Omega_M^{(1,1)} & \Omega_M^{(1,0)} \\ \Omega_M^{(0,1)} & \Omega_M^{(0,0)} \end{pmatrix}.$$

The value of Ω_2 can be calculated directly; it is

$$\Omega_2 = \begin{pmatrix} e^{-2\alpha+\beta} & e^{-\alpha} \\ e^{-\alpha} & 1 \end{pmatrix}.$$

Moreover, it is not hard to show the correct update equation is $\Omega_{M+1} = A\Omega_M$, where

$$A = \begin{pmatrix} e^{-\alpha+\beta} & e^{-\alpha} \\ 1 & 1 \end{pmatrix}.$$

If site $M + 1$ is an encouraged site, then we need to use a slightly different update matrix, in which $\alpha - \gamma$ is substituted for α:

$$A_{\text{encourage}} = \begin{pmatrix} e^{\gamma-\alpha+\beta} & e^{\gamma-\alpha} \\ 1 & 1 \end{pmatrix}.$$

Finally, the partition function for a *closed* one-dimensional MRF is easily seen to be

$$Z_M = \mathrm{Tr}\left[\begin{pmatrix} e^{\beta} & 1 \\ 1 & 1 \end{pmatrix}\Omega_M\right].$$

5.10 Further remarks

Occlusion is a perennial problem for anyone working in object localisation. This chapter introduced a new way of obtaining realistic inferences about partially occluded targets. Specifically, it addressed the problem of how to adjust the generative models of Chapter 3 so that the likelihoods of occluded targets are not over-penalised. This was done by removing the assumption that occlusion events on different measurement lines are independent: instead, the measurement lines are treated as sites in an MRF whose behaviour is chosen to represent the types of occlusion expected in practice. Using this MRF as the prior for a Bayesian approach, a new likelihood termed the Markov likelihood was derived.

Two situations were demonstrated in which the Markov likelihood produces far more realistic output than the previously introduced independence likelihood. Firstly, it was shown that if the Markov likelihood is used to analyse a scene with one unoccluded target and one partially occluded target, the two peaks in the posterior distribution of target configurations have similar magnitudes (see Figure 5.5). This is in marked contrast to an identical analysis using the independence likelihood, where the posterior peaks differ by several orders of magnitude. Secondly, it was shown the Markov likelihood can significantly improve differentiation between partially occluded targets and background clutter (Figure 5.8). This is because the Markov likelihood takes account of which types of occlusion have good support from the prior.

The success of the Markov likelihood shows it is possible to apply rigorous probabilistic modelling and Bayesian methods to the classic problem of occlusion. Moreover, it represents a significant advance in the potential of recognition systems to provide meaningful statistical information to higher-level systems. It remains to be seen, however, whether any of the current recognition paradigms can realise this potential.

6

A probabilistic exclusion principle for multiple objects

6.1 Introduction

This chapter proposes a mathematically rigorous methodology for tracking multiple objects when the number of objects is fixed in advance. The fundamental problem to be addressed is demonstrated in Figure 6.1. Two instantiations of the same tracking algorithm, with different initial conditions, are used to track two targets simultaneously. When one target passes close to the other, both tracking algorithms are attracted to the single target which best fits the head-and-shoulders model being used. One might think of avoiding this problem in a number of ways: interpreting the targets as "blobs" which merge and split again (Haritaoglu et al., 1998; Intille et al., 1997), enforcing a minimum separation between targets (Rasmussen and Hager, 1998), or incorporating enough 3D geometrical information to distinguish the targets (Koller et al., 1994). However, each of these solutions can be unattractive. A blob interpretation does not maintain the identity of the targets, and is difficult to implement for moving backgrounds and for targets which are not easily segmented. A minimum separation relies on heuristics and fails if the targets overlap. Incorporating 3D information is impossible without detailed scene modelling.

So it seems we must instead address the fundamental problem: that the observation model used to interpret image measurements permits two targets to occupy the same point in configuration space too easily. More specifically, a single piece of image data (such as an edgel, or a colour blob) must not simultaneously reinforce mutually exclusive hypotheses. What is needed is a "probabilistic exclusion principle", and an observation model exhibiting this behaviour is described in this chapter. The formal model will initially be derived for "wire-frame" targets — objects which have detectable boundaries but which do not occlude each other. We then describe how occlusion reasoning about solid objects can be incorporated naturally into the same framework. The most interesting feature of this approach is that it works even when the

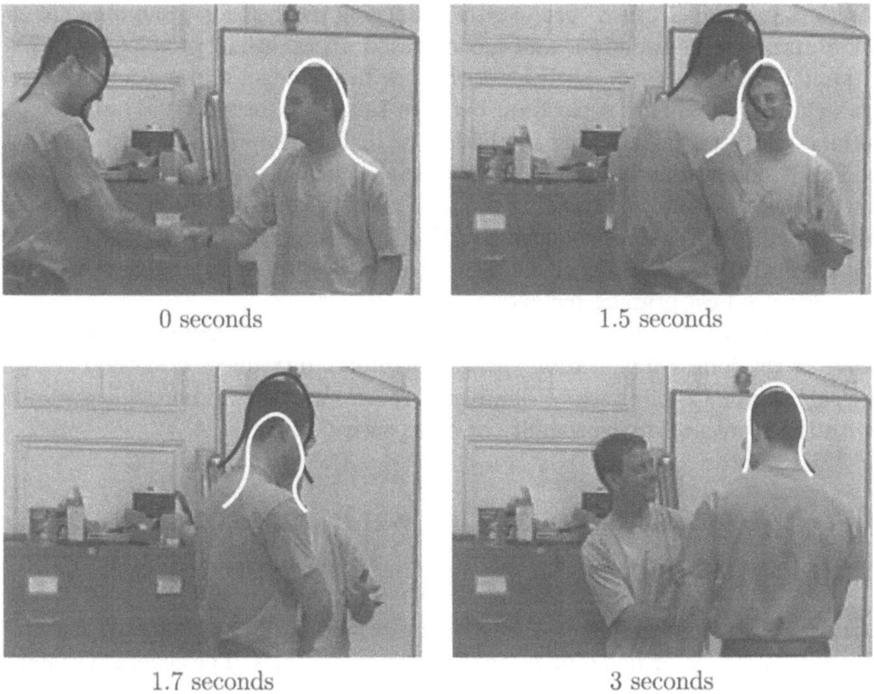

Figure 6.1: ***With an observation model designed for one target, two trackers initialised in distinct configurations eventually lock on to the one target which bests fits the model.*** *The objective is to derive an observation model which does not permit the presence of two targets to be inferred from measurements of only one. This sequence can be viewed as a movie from the book web site (see page 3).*

targets are *indistinguishable given the available information.* This is of both theoretical and practical interest.

Many visual tracking systems for multiple objects have been developed. One standard technique is the probabilistic data association filter (PDAF) (Bar-Shalom and Fortmann, 1988), and other successful examples include (Haritaoglu et al., 1998; Intille et al., 1997; Paragios and Deriche, 1998; Rasmussen and Hager, 1998). These generally employ a combination of blob identification and background subtraction; both techniques are complementary to the method proposed here. In particular, the exclusion principle does not allow two targets to merge when their configurations become similar; instead, the model continues to interpret the data in terms of two targets. As will be seen, it is a natural consequence of the methodology that the probability distribution for an obscured target diffuses until it is reinforced by further data. Furthermore, the method works for unknown and constantly changing backgrounds. Rasmussen and Hager (1998) proposed a promising method for combining colour blob and edge information, and incorporated an exclusion principle by using a joint PDAF. However, their algorithm for fusing edgel information enforced an arbitrary minimum separation between targets. Gordon (1997) employs a similar multi-target tracking methodology to this chapter but with a rather different observation model and no explicit exclusion principle. Probably the most careful system for tracking multiple targets while maintaining exclusion is the multiple hypothesis tracking of Cox and Hingorani (1996), which builds on Reid (1979). This algorithm is motivated by probabilistic reasoning, but involves a k-best pruning technique.

One of the difficulties with tracking multiple objects is the high dimensionality of the joint configuration space. Chapter 7 introduces a method known as *partitioned sampling* which diminishes the computational burden associated with the increased dimensionality of multi-target spaces.

6.2 A generative model with an exclusion principle

As usual, the measurements $\mathbf{Z}$ of the image will consist of detected features $(\mathbf{z}^{(1)}, \ldots \mathbf{z}^{(M)})$ on M one-dimensional measurement lines. The generative model need only specify how these features arise; it will be a relatively simple generalisation of the generic model of Section 3.1. In fact, all we need to do is replace Assumptions 1 and 2 with analogues which make sense for multiple targets.

The intersection number c and intersection innovation $\boldsymbol{\nu}$ are defined in the same way as in Section 3.1.1 — but now, of course, the case $c \geq 2$ will not be neglected. Figure 6.2 gives a concrete example of how to calculate c and $\boldsymbol{\nu}$ when there are two targets.

6.2.1 Description of the generative model

Suppose that a given measurement line of length L has intersection number c and intersection innovation $\boldsymbol{\nu} = (\nu_1, \ldots \nu_c)$ with respect to a configuration $\mathbf{x}$. Ignore, for the moment, the possibility that the feature detector might not detect each of the intersections. Then each intersection $\nu_i, i = 1, \ldots c$, will correspond to a detected feature z_i, say. The generative model replaces Assumption 1 (page 39) by

- **Assumption 1′** Each z_i is drawn randomly and independently from the boundary density function $\mathcal{G}(z_i|\nu_i)$. Thus the pdf for $(z_1, \ldots z_c)$ is

$$\prod_{i=1}^{c} \mathcal{G}(z_i|\nu_i).$$

As in Section 3.1, the boundary density function is taken to be a truncated Gaussian.

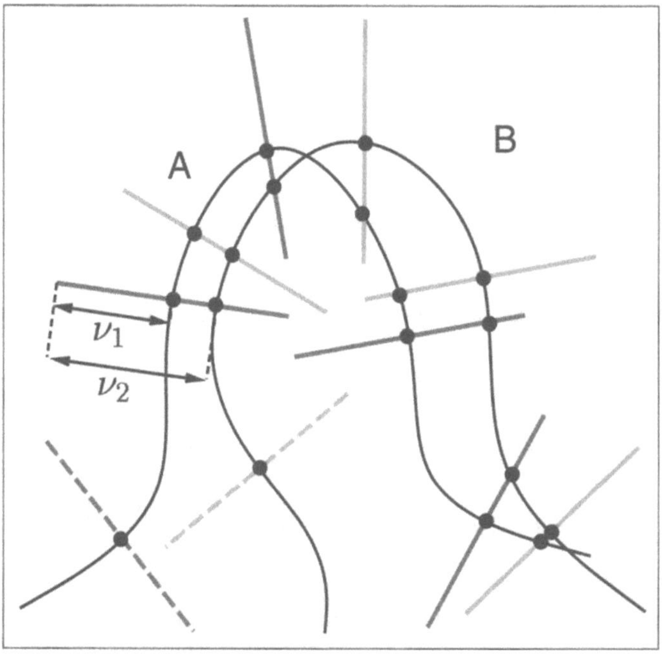

Figure 6.2: ***Intersection numbers for multiple targets.*** *The solid curves show hypothesised outlines of two head-and-shoulders targets, A, and B. Intersections for several measurement lines are shown as solid circles. The solid lines have intersection number* $c = 2$*, and the dashed lines have* $c = 1$*. For one of the measurement lines, the intersection innovation* $\boldsymbol{\nu} = (\nu_1, \nu_2)$ *is shown.*

It is important to allow for the possibility that one or more of the boundary features is not detected. This is expressed by replacing Assumption 2 (page 39) with an obvious analogue:

- **Assumption 2′** For each $c \geq 1$, and each $i = 0, 1, \ldots c$, there is a fixed probability q_{ic} that only i of the c boundaries are detected. The set of which boundaries are not detected is selected uniformly at random from the $\binom{c}{i}$ possibilities. Note that $\sum_{i=0}^{c} q_{ic} = 1$.

The remaining assumptions of the original generative model (Assumptions 3–6, Chapter 3) are unchanged.

6.2.2 Likelihoods derived from the generative model

The new set of assumptions leads to an analogue of Proposition 3.1, giving the pdfs for general intersection numbers c.

Proposition 6.1 *Suppose the non-detection probabilities are zero (i.e.* $q_{ic} = 0$ *for* $i = 0, 1, \ldots, (c-1)$ *and* $q_{cc} = 1$*). Then the pdf for the generative model on* *a single measurement line of length L is given by*

$$p_c(n; \mathbf{z}|\boldsymbol{\nu}) = \frac{b(n-c)}{L^{n-c}\binom{n}{c}} \sum_{i_1<i_2<\ldots<i_c} \prod_{j=1}^{c} \mathcal{G}(z_{i_j}|\nu_j). \tag{6.1}$$

Proof. This can be proved by the same arguments as for the proof of Proposition 3.1, page 42. ■

Of particular interest are the results for $c = 0, 1, 2$:

$$p_0(n; \mathbf{z}) = b(n)/L^n \tag{6.2}$$

$$p_1(n; \mathbf{z}|\boldsymbol{\nu} = \{\nu_1\}) = b(n-1) \sum_{k=1}^{n} \mathcal{G}(z_k|\nu_1)/nL^{n-1} \tag{6.3}$$

$$p_2(n; \mathbf{z}|\boldsymbol{\nu} = \{\nu_1, \nu_2\}) = b(n-2) \sum_{i \neq j} \frac{\mathcal{G}(z_i|\nu_1)\mathcal{G}(z_j|\nu_2)}{L^{n-2}n(n-1)}. \tag{6.4}$$

Note that p_0 and p_1 are exactly the same as in Proposition 3.1, as they should be since our new Assumptions $1'$ and $2'$ reduce to the original ones for $c \leq 1$.

The new assumptions also lead to an analogue of Proposition 3.2, which expresses the pdfs for non-zero values of the q_{ic}:

Proposition 6.2 *The pdfs for the generative model on a single measurement line of length L are given by*

$$\tilde{p}_c(n; \mathbf{z}|\boldsymbol{\nu}) = \sum_{i=1}^{c} q_{ic} p_i(n; \mathbf{z}|\boldsymbol{\nu}). \tag{6.5}$$

Proof. This is a trivial consequence of Assumption $2'$. ■

Typical graphs of $\tilde{p}_1$ and $\tilde{p}_2$ are shown in Figure 6.3.

6.2.3 Where does the "exclusion principle" come from?

As you can see from the bottom panel of Figure 6.3, the likelihood function $\tilde{p}_2(\mathbf{z}|\nu_1, \nu_2)$ incorporates an exclusion principle: the peaks on the line $\nu_1 = \nu_2$ are much smaller than peaks off the line, so the likelihood of both targets having the same innovation is much lower. Mathematically, this exclusion principle arises from the fact that the sum (equation 6.4) defining p_2 does not include the $i = j$ terms — we *exclude* the possibility that a single z_i corresponds to two different intersections. The reason for this stems from the generative model: each boundary is assumed to generate its own boundary feature, independently of the others.

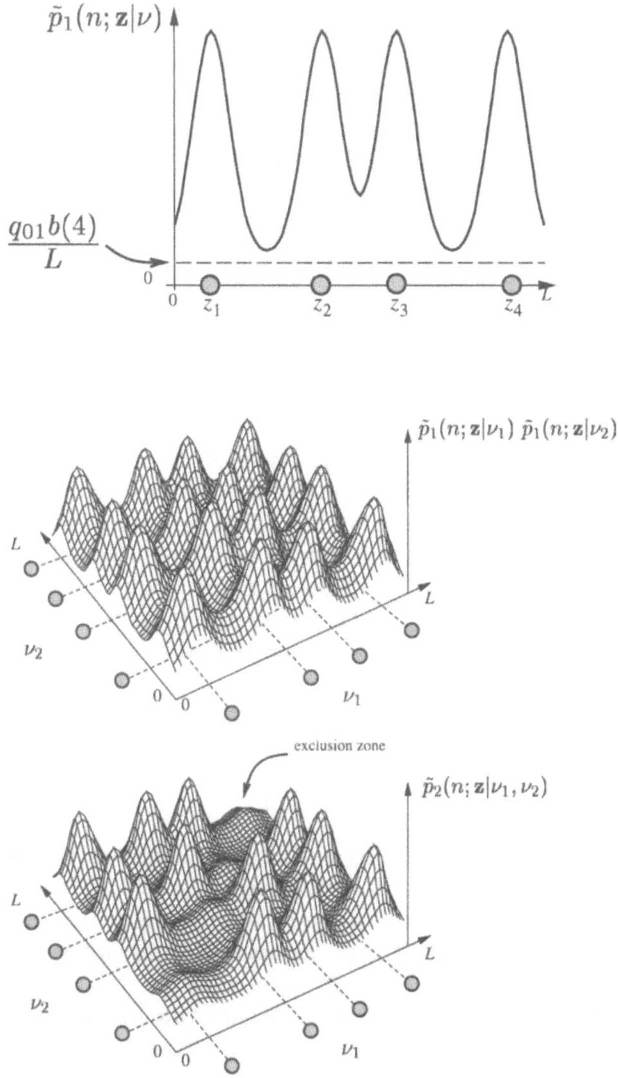

Figure 6.3: ***One-target and two-target likelihood functions for a single measurement line.*** *Top: the one-target likelihood function $\tilde{p}_1$ (see also Figure 3.3). The likelihood is a linear combination of shifted copies of $\mathcal{G}(z|\cdot)$ and of the constant p_0. It peaks near the four measurements z_i (shown as shaded circles). Middle: A naïve equation -target likelihood $\tilde{p}_1(n; \mathbf{z}|\nu_1)\tilde{p}_1(n; \mathbf{z}|\nu_2)$ formed by taking the product of two one-target densities (top panel). The likelihood peaks near pairs of measurements z_i, z_j (shaded circles and dotted lines). Bottom: the two-target likelihood $\tilde{p}_2(n; \mathbf{z}|\nu_1, \nu_2)$ derived from the generative model. Again, the likelihood peaks near pairs of measurements z_i, z_j (shaded circles and dotted lines), but now a probabilistic exclusion principle operates: because the sum (equation 6.4) in the definition of p_2 excludes $i = j$, the probability peaks are much smaller on the line $\nu_1 = \nu_2$.*

6.2.4 The full likelihood

The above discussion was framed in terms of a single measurement line, but, for any given hypothesised configuration $\mathbf{x}$, the measurements $\mathbf{Z}$ will arise from say M distinct measurement lines. Invoking Assumption 6 (independence of measurement lines), the *exclusion likelihood function* is defined in just the same way as in Proposition 3.3 (page 44):

$$\mathcal{P}_{\text{excl}}(\mathbf{Z}|\mathbf{x}) = \prod_{i=1}^{M} \tilde{p}_{c_{\mathbf{x}}(i)}(\mathbf{z}^{(i)}|\boldsymbol{\nu}_{\mathbf{x}}(i)). \tag{6.6}$$

6.3 Tracking multiple wire-frame objects

As in previous chapters, tracking is performed here using the Condensation algorithm, which is capable of dealing with complex likelihood functions such as equation 6.6. Recall from Chapter 2 that a Condensation tracker represents the state of a system at time t by a particle set S^t which represents the posterior distribution $p(\mathbf{x}^t|\mathcal{Z}^t)$. Here $\mathbf{x}^t$ is the configuration at time t, and $\mathcal{Z}^t = (\mathbf{Z}^1, \ldots \mathbf{Z}^t)$ is the history of all measurements up to time t. At each time-step the current particle set S^t is propagated according to the Condensation diagram introduced in Chapter 2:

$$\underset{(a)}{\boxed{S^t}} \longrightarrow \boxed{\sim} \longrightarrow \left\langle * p(\mathbf{x}'|\mathbf{x}) \right\rangle \underset{(b)}{\longrightarrow} \left\langle \times \mathcal{P}(\mathbf{Z}^{t+1}|\mathbf{x}') \right\rangle \longrightarrow \underset{(c)}{\boxed{S^{t+1}}} \tag{6.7}$$

where $p(\mathbf{x}'|\mathbf{x})$ is a conditional distribution modelling the system dynamics and $\mathcal{P}(\mathbf{Z}^{t+1}|\mathbf{x}')$ is the likelihood of the measurements $\mathbf{Z}^{t+1}$. The $\sim$ symbol represents resampling, the $*$ is application of the dynamics and the $\times$ represents multiplication (i.e. re-weighting) by the measurement density; all these operations were described in Chapter 2. The labels (a)–(c) refer to an example given later, in Figure 7.4. Of course, to demonstrate the exclusion principle we use the likelihood function $\mathcal{P}_{\text{excl}}(\mathbf{Z}|\mathbf{x})$ as the measurement density. Note that $\mathcal{P}_{\text{excl}}$ as defined in equation 6.6 is not valid for opaque objects, since the model expects to observe all boundaries, even when occluded. However, it is valid for wire-frame objects, so an experiment on wire frames was performed. As a control for the experiment, we need a likelihood $\mathcal{P}'$, similar to $\mathcal{P}_{\text{excl}}$, but which does not incorporate an exclusion principle. Naming the two targets A and B, and writing $c_A(i)$ for the number of intersections of A with line i, let $\nu_A^{(i)}$ be the coordinates of these intersections and define the *one-body density*

$$\mathcal{P}_A(\mathbf{Z}|\mathbf{x}) = \prod_{i=1}^{M} \tilde{p}_{c_A(i)}(\mathbf{z}^{(i)}|\boldsymbol{\nu}_A^{(i)}), \tag{6.8}$$

and similarly for $\mathcal{P}_B$. We take $\mathcal{P}' = \mathcal{P}_A\mathcal{P}_B$, so the posteriors for A and B given $\mathbf{Z}$ are treated as independent. A typical graph of $\mathcal{P}'$ for just one measurement line is shown as the middle panel of Figure 6.3. Note that this is the same as for $\mathcal{P}_{\text{excl}}$ (the lower panel), but with four additional peaks down the line $\nu_1 = \nu_2$. Figure 6.4 shows the tracking results: as expected, $\mathcal{P}_{\text{excl}}$ successfully maintains exclusion between the targets whereas $\mathcal{P}'$ does not.

6.4 Tracking multiple opaque objects

The wire-frame model can be adapted for use with solid objects. The method uses the mixed state Condensation tracker of Isard and Blake (1998c), combined with a "2.1D" model of the targets. The basic idea of a mixed state Condensation tracker is that each particle carries a discrete label in addition to the continuous parameters describing its configuration. The extended state is defined to be

$$\mathbf{X} = (\mathbf{x}, y), \mathbf{x} \in \mathcal{X}, y \in \{1, \ldots N_s\}, \tag{6.9}$$

where y is a discrete variable labelling the current model, and $\mathbf{x}$ is a vector of continuous parameters specifying the configuration of the targets. In the two-object case, $\mathbf{x} = (\mathbf{x}^A, \mathbf{x}^B)$ and y can take one of two values: $y = 1$ if A is nearer the camera than B, and $y = 2$ if B is nearer than A. This is what we mean by a 2.1D model: the only 3D geometric aspect to be inferred is whether target A can occlude target B or vice versa. The idea of 2.1D models is now a common one in computer vision: introduced via the "2.1D sketch" by Mumford and Nitzberg (1990), the concept has evolved into "layers" used for tracking and mosaicking (e.g. Irani and Anandan, 1998).

The process density can then be decomposed as follows:

$$p(\mathbf{X}_t|\mathbf{X}_{t-1}) = P(y_t|\mathbf{x}_t, \mathbf{X}_{t-1})\, p(\mathbf{x}_t|\mathbf{x}_{t-1}),$$

and if $y_{t-1} = j$ and $y_t = i$ this can be written more explicitly as

$$p(\mathbf{X}_t|\mathbf{X}_{t-1}) = T_{ij}(\mathbf{x}_t, \mathbf{x}_{t-1}) p(\mathbf{x}_t|\mathbf{x}_{t-1}).$$

where T_{ij} is a transition matrix and p is a density specifying the continuous dynamics for a particle. Here it is appropriate for $T_{ij}(\mathbf{x}_t, \mathbf{x}_{t-1})$ to be independent of $\mathbf{x}_{t-1}$. If $\mathbf{x}_t^A$ and $\mathbf{x}_t^B$ overlap then the occlusion relationship cannot change in the current time-step and so we take $T_{ij}(\mathbf{x}_t)$ to be the identity matrix. If $\mathbf{x}_t^A$ and $\mathbf{x}_t^B$ do not overlap then there is a small, fixed probability that y will change, represented by taking $T_{ij}(\mathbf{x}_t) = \begin{pmatrix} 1-\delta & \delta \\ \delta & 1-\delta \end{pmatrix}$ with $0 < \delta \ll 1$.

The mixed state Condensation tracker presented here incorporates a significant difference to that of Isard and Blake (1998c) — the observation density $p(\mathbf{Z}_t|\mathbf{X}_t)$ depends not only on $\mathbf{x}_t$ but also on the discrete state y_t. The multi-target likelihood function equation 6.6 is used, but now the intersection counts $c_{\mathbf{x}}(i)$ are calculated using the discrete variable y and the 2.1D geometry to determine if a given boundary feature should be visible or not, as in Figure 6.5. To emphasise this we can write $c_{\mathbf{x}}(i, y)$ for the number of *visible* target boundaries intersecting the ith measurement line of a configuration $(\mathbf{x}, y)$; the

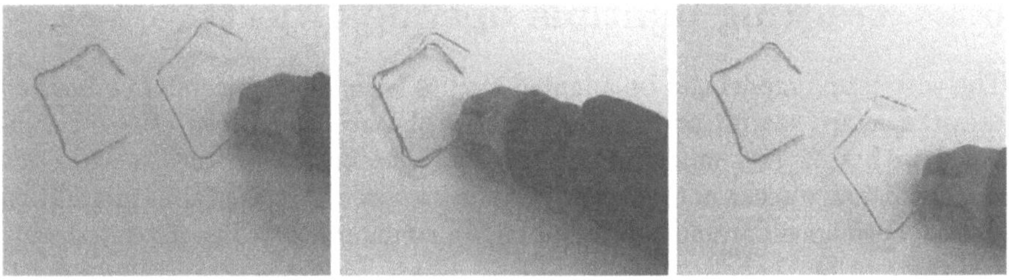

(a) *raw sequence of moving wire frames*

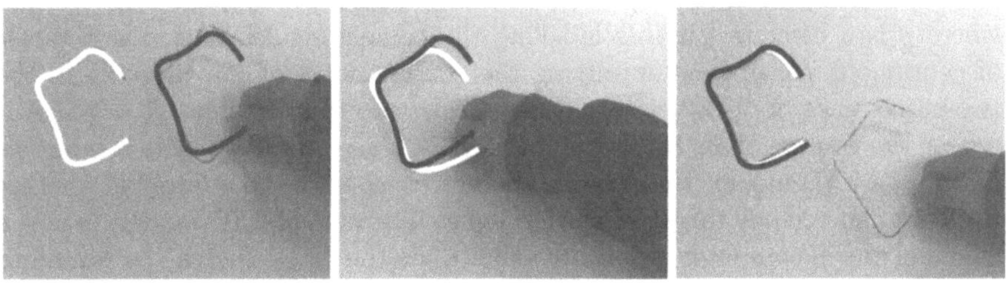

(b) *tracking using product of two one-target observation densities*

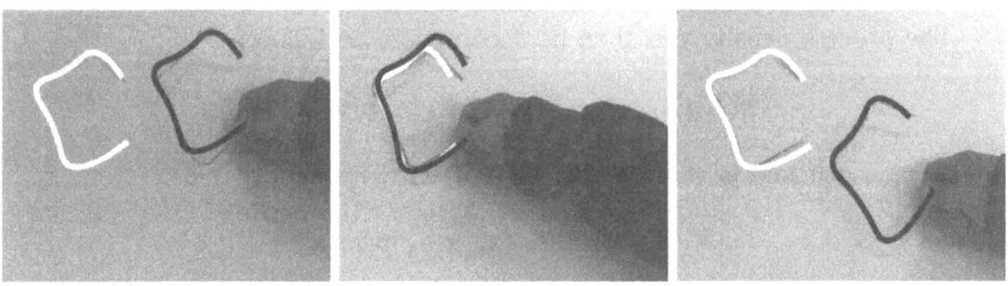

(c) *tracking using probabilistic exclusion principle*

Figure 6.4: ***The exclusion principle operating on a wire-frame example.*** *(a) Three stills from a sequence of two pieces of wire with similar shapes. Note that for several frames in the middle of the sequence, the two wires have very similar configurations. (b) Results using the likelihood $\mathcal{P}'$, which does not incorporate an exclusion principle. When the configurations become similar, both targets settle on the best-fitting wire. (c) Successful tracking using the exclusion principle likelihood $\mathcal{P}_{\text{excl}}$.*

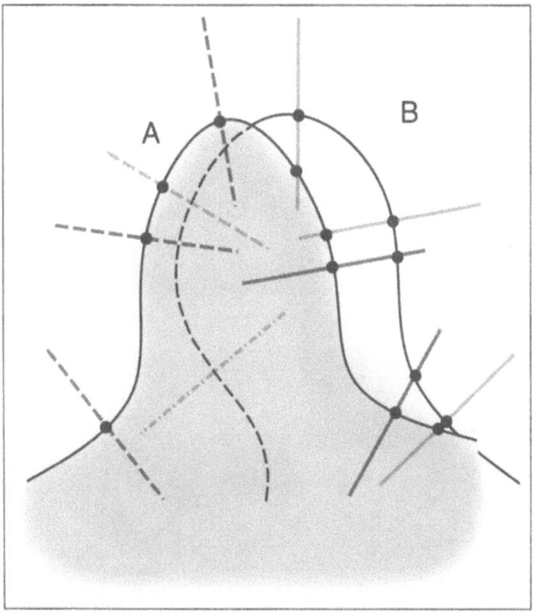

Figure 6.5: ***Intersection numbers calculated from 2.1D geometry.*** *In this diagram, $y = 1$, meaning the shaded area is occluded by target A. Visible intersections of measurement lines and target boundaries are shown as solid circles. The solid lines have intersection number $c = 2$, dotted have $c = 1$ and dashed $c = 0$. These are the c-values used in equation 6.10.*

coordinates of the visible boundaries on the ith line are written $\boldsymbol{\nu}_{\mathbf{x}}(i, y)$. Then the likelihood in the occluded case is

$$\mathcal{P}_{\text{occl}}(\mathbf{Z}|\mathbf{x}) = \prod_{i=1}^{M} \tilde{p}_{c_{\mathbf{x}}(i,y)}(\mathbf{z}^{(i)}|\boldsymbol{\nu}_{\mathbf{x}}(i, y)). \tag{6.10}$$

To understand this, compare with equation 6.6. The functions $\tilde{p}_c, c = 0, 1, 2,$ are still as defined in equations 6.2–6.4. The only change is that the intersection numbers c and target boundary positions ν now depend on the discrete state y which specifies which target is in front of the other. The derivation of equation 6.10 is otherwise identical to equation 6.6. A detailed example is given in Figure 6.5.

The likelihood $\mathcal{P}_{\text{occl}}$ performs well in experiments. Figure 6.6 shows a typical sequence involving occlusion. The configuration space has 16 dimensions: eight key-frames from principal components analysis of templates (Baumberg and Hogg, 1994; Cootes and Taylor, 1992), for each of two targets. Tracking is performed with $n = 2000$ particles, and predictive dynamics in the form of Brownian motion with an amplitude matched to the speed of a walking person. Note how the occluded contours diffuse at 0.7 seconds. Because of the exclusion principle they coalesce again only when some evidence from the correct

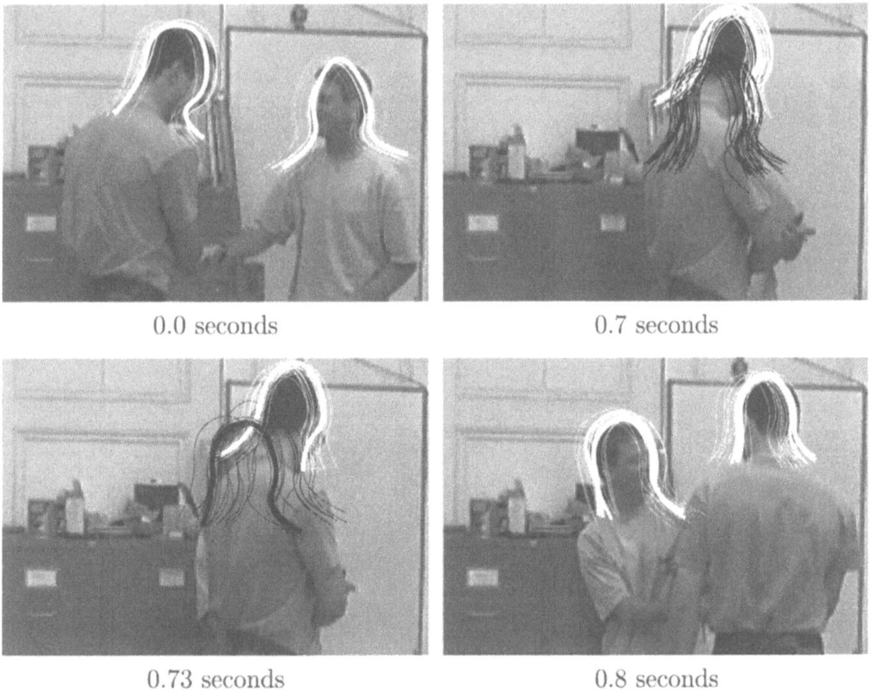

Figure 6.6: ***Correct tracking with a density incorporating occlusion reasoning (see also Figure 6.1).*** *20 of the 2000 particles are shown in each frame, with widths proportional to their probabilities. Initially, each particle consists of two white contours: one initialised on each of the two targets. A contour is drawn in black if its value of y, as defined in equation 6.9, implies that it is partially occluded. This sequence can be viewed as a movie from the book web site (see page 3).*

target is observed. The undesirable tracking behaviour of Figure 6.1 has been corrected.

As a canonical tracking challenge, the same multiple target methodology was applied to the "leaf sequence" used in Isard and Blake (1998c). Two leaves were tracked, using an affine shape space and $N = 4000$ samples with learnt dynamics. The need for 4000 samples is reduced to 750 by the partitioned sampling method described in the next chapter. Tracking is successful despite occlusions; some stills are shown in Figure 6.7.

Figure 6.8 gives details of the parameter values used for all the experiments.

Figure 6.7: ***Tracking multiple leaves, in moving clutter and with occlusions.*** *Three stills from a tracked sequence are shown. The black contour shows a correctly inferred occlusion.*

Non-detection probabilities, $c = 1$	(q_{01}, q_{11})	$(0.1, 0.9)$
Non-detection probabilities, $c = 2$	(q_{02}, q_{12}, q_{22})	$(0.05, 0.2, 0.75)$
Clutter feature probabilities	$b(n)$	MLE from first frame of sequence
Discrete transition probability	δ	0.01
Boundary feature distribution	$\mathcal{G}(z\|\nu)$	Gaussian with std dev of 7 pixels
Length of measurement lines	L	40 pixels

Figure 6.8: ***Parameter values and other choices used for experiments.*** *The non-detection probabilities were determined by trial and error on simple examples. The discrete transition probability corresponds to a time constant of 2.0 seconds for a given discrete state. The standard deviation of the boundary feature distribution is estimated from templates fitted to the targets. The measurement lines extend approximately three of these standard deviations in each direction.*

7

Partitioned sampling

A potential limitation of the Condensation algorithm is that if the state space has many dimensions, then the number of particles required to model a distribution can be very large indeed. This is of particular concern when tracking multiple objects, since the number of dimensions in the state space is proportional to the number of objects. Fortunately, "partitioned sampling" significantly reduces this curse of dimensionality. It is the statistical analogue of a hierarchical search: the intuition is that it should be more efficient to search *first* for whichever target is unoccluded, and only then to search for another target which may lie behind.

7.1 The need for partitioned sampling

Why does the Condensation algorithm require more particles when there are more targets? This section attempts to answer this question; it involves only simple ideas but they look complicated when formalised, so it will be better to have an informal overview first:

> **Informal overview of this section:** for a given implementation of the Condensation algorithm, the survival diagnostic $\mathcal{D}(n)$ is approximately αn for some constant $\alpha < 1$ (called the *survival rate*) when n is large. It turns out that the number of particles required for a fixed level of tracking performance is given by K/α for some constant K. Suppose the same implementation of Condensation is now used to track d objects rather than just one object. Then it turns out the number of particles required is K/α^d. Since in most practical contour tracking tasks, $\alpha \ll 1$, the number of particles required for correct tracking of multiple objects becomes unfeasibly large. Partitioned sampling is a way of avoiding this problem.

These comments will now be made a little more rigorous. First, let us see why $\mathcal{D}(n) \approx \alpha n$. Recall the scenario of Section 2.6.1: a particle set S has been formed with prior (or "proposal density") $q(\mathbf{x})$ and weighted by $p(\mathbf{x})/q(\mathbf{x})$, resulting in a posterior $p(\mathbf{x})$. Some simple calculations give

$$\begin{aligned}
\mathcal{D}(n) &= \left(\textstyle\sum_{i=1}^{n} \pi_i^{(n)2}\right)^{-1} && \text{by definition of } \mathcal{D}\text{, equation 2.19} \\
&\approx \left(\textstyle\sum_{i=1}^{n} \frac{1}{n^2} p(\mathbf{x}_i^{(n)})^2 / q(\mathbf{x}_i^{(n)})^2\right)^{-1} && \text{by definition of the } \pi_i\text{, and Lemma 2.7 (page 33)} \\
&\approx \left(\tfrac{1}{n} \textstyle\int p(\mathbf{x})^2/q(\mathbf{x})\, d\mathbf{x}\right)^{-1} && \text{as the } \mathbf{x}_i^{(n)} \text{ are drawn from } q(\mathbf{x}) \\
&= \left(\textstyle\int p(\mathbf{x})^2/q(\mathbf{x})\, d\mathbf{x}\right)^{-1} \times n.
\end{aligned}$$

So we do indeed have $\mathcal{D}(n) \approx \alpha n$, provided the following definition is adopted.

Definition (Survival rate) When performing factored sampling with proposal density q and posterior p, the *survival rate* is defined to be

$$\alpha = \left(\int p(\mathbf{x})^2/q(\mathbf{x})\, d\mathbf{x}\right)^{-1}. \tag{7.1}$$

(See theorem 2 of Geweke (1989) for another use of this quantity.) Note that since this can be written as $(\int p/q\,(p\,d\mathbf{x}))^{-1}$, we can think of $1/\alpha$ as the expected value of p/q, where the expectation is taken with respect to the posterior p. There are two important special cases where α can be calculated explicitly. Firstly, suppose q is a uniform distribution on a set $\mathcal{X}_q \subset \mathcal{X}$ of volume V_q, and that p is also uniform, on a smaller subset $\mathcal{X}_p \subset \mathcal{X}_q$ of volume V_p. Then p/q is equal to V_q/V_p everywhere on $\mathcal{X}_p$, so that

$$\alpha = V_p/V_q.$$

That is, the survival rate is just the ratio of the volume of the posterior to the volume of the prior.

The other important special case is when q is a broad Gaussian of variance σ_q^2 centred on zero, and p is a narrower Gaussian of variance σ_p^2 centred on t, say. Then direct calculation shows that with $\sigma = \sigma_p/\sigma_q$,

$$\begin{aligned}
\alpha &= \exp\left(\frac{-t^2}{2-\sigma^2}\right)\sigma^2\sqrt{2/\sigma^2 - 1} \\
&= \sqrt{2}e^{-t^2/2}\frac{\sigma_p}{\sigma_q} + O\left((\sigma_p/\sigma_q)^3\right).
\end{aligned}$$

Again, the survival rate is related to the ratio of the volume of the posterior to volume of prior, but this time the factor $e^{-t^2/2}$ is present to account for the amount of the posterior's volume which lies "inside" the prior's typical volume.

Example Figure 7.1 shows an example of a survival rate α calculated for a contour likelihood in a real image. In this particular example, in which the configuration space is the one-dimensional interval $[-150, 150]$, the value of α was calculated numerically as 0.20.

As was argued in Section 2.6.1 (page 28), one sensible way of defining the "acceptable performance" of a tracker is to ensure that $\mathcal{D}(n) > n_{\min}$, for some

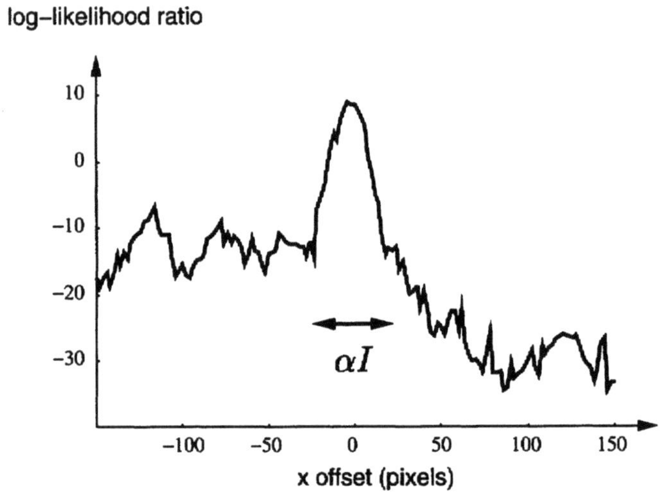

Figure 7.1: ***Survival rate.*** *One of the likelihoods of Figure 3.4 (page 53) is shown for a larger range of x-offsets. (The contour and image from which these likelihoods are calculated are in Figure 3.5(b).) Taking a uniform prior q on the interval I shown, the survival rate α for this particular likelihood function can be calculated numerically as 0.20. This corresponds to the "volume" αI indicated on the graph.*

minimum acceptable effective sample size $n_{\min}$. Since $\mathcal{D}(n) \approx \alpha n$, this implies that we must take $n > n_{\min}/\alpha$.

Finally, suppose the same algorithm is now used to perform factored sampling on a Cartesian product of d copies of the configuration space $\mathcal{X}$ (this is exactly what it means to track d objects using "the same implementation"). Then this results in a new survival rate α' calculated from $p' = p^d$ and $q' = q^d$. It is immediate from equation 7.1 that $\alpha' = \alpha^d$ and that, therefore, we must now take $n > n_{\min}/\alpha^d$ to obtain the same level of tracking performance.

The intuition that α is the ratio of the posterior and prior volumes gives a hint as to how this problem could be solved. Take the simple case of tracking two objects A and B, whose configurations are described respectively by the one-dimensional variables $x_A, x_B \in [0, 1]$. Suppose the survival rate for the one-object problem is α: then for the two-object problem, we have a survival rate $\alpha' = \alpha^2$. Figure 7.2 shows a schematic representation of the situation. The intuition behind partitioned sampling is that, instead of searching the entire unit square for the lightly shaded area α', we can divide the search into two stages: first, a search of the horizontal axis only, which will attempt to populate the dark shaded area α. This step will have survival rate α. Second, we try to populate the lightly shaded area. This second step will also have survival rate of approximately α, since the *relative* area of the dark shade to light shade is α'/α. This is the key idea behind partitioned sampling. It remains to show

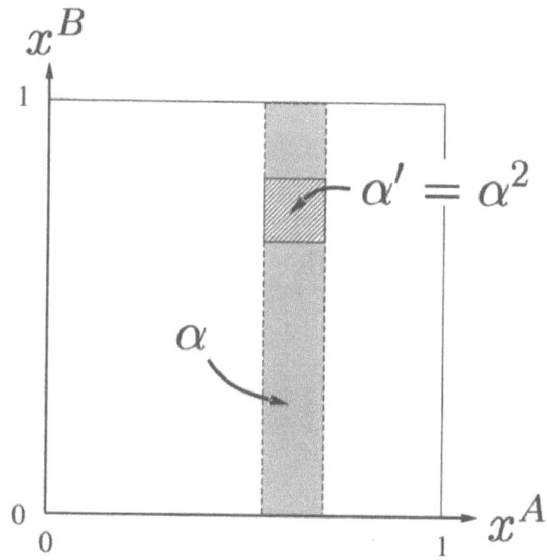

Figure 7.2: ***Intuition behind partitioned sampling.*** *To locate the peak of a 2D likelihood function, which has area $\alpha' = \alpha^2$, the search is split into two stages, each of which has survival rate α. The first stage populates the shaded area with particles, and the second stage populates the hatched area.*

how we can "populate" certain parts of the configuration space with particles in the desired manner. This is done using a new operation called weighted resampling.

7.2 Weighted resampling

The partitioned sampling algorithm requires an additional operation on particle sets, termed weighted resampling. Like the standard resampling operation, weighted resampling does not alter the distribution represented by the particle set. However, it can be used to reposition the locations of the particles so that the representation is more efficient for future operations.

Definition (Weighted resampling) Let $S = (\mathbf{x}_i^{(n)}, \pi_i^{(n)})_{i=1}^n, n = 1, 2, \ldots,$ be a particle set, and let g be a strictly positive function on the configuration space $\mathcal{X}$. The *resampling weights* for g are random variables $\rho_j^{(n)}, j = 1, 2, \ldots n$, $n = 1, 2, \ldots,$ defined by

$$\rho_j^{(n)} = \frac{g(\mathbf{x}_j^{(n)})}{\sum_{i=1}^n g(\mathbf{x}_i^{(n)})}.$$

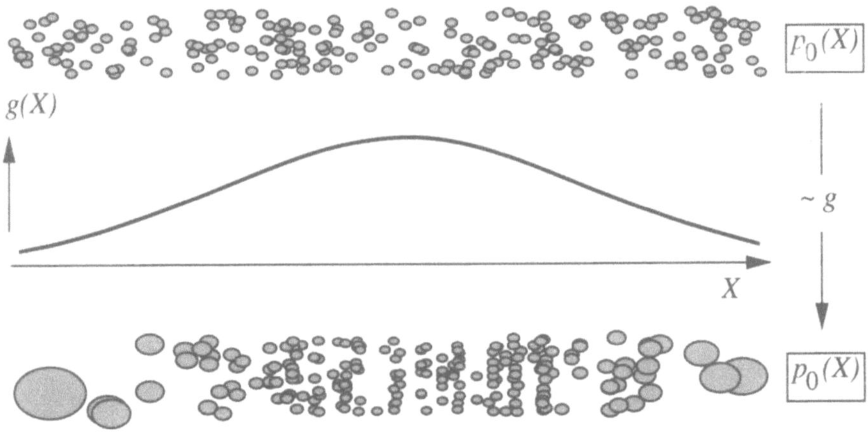

Figure 7.3: ***Weighted resampling.*** *A uniform prior $p_0(X)$, represented as a particle set (top), is resampled via an importance resampling function g to give a new, reweighted particle set representation of p_0. Note that these are one-dimensional distributions; the particles are spread in the y-direction only so they can be seen more easily.*

Define a new particle set, S' by

$$\mathbf{x}_i^{(n)'} = \mathbf{x}_j^{(n)} \text{ with probability } \rho_j^{(n)}$$
$$\pi_i^{(n)'} \propto \pi_j^{(n)} / \rho_j^{(n)},$$

where the random choice of $\mathbf{x}_i^{(n)'}$ occurs independently for $i = 1, \ldots n$. We say S' has been obtained from S by *weighted resampling with respect to the importance function* g, and depict this operation on a Condensation diagram by

$$\underset{S}{\boxed{p(\mathbf{x})}} \longrightarrow \boxed{\sim g} \longrightarrow \underset{S'}{\boxed{p(\mathbf{x})}} \tag{7.2}$$

Intuitively, $g(\mathbf{x})$ is a function with high values in regions where we would like to have many particles. The objective of the weighted resampling is to populate such regions so that subsequent operations on the particle set will produce accurate representations of the desired probability distributions.

Figure 7.3 shows a simple one-dimensional example of weighted resampling with respect to an importance function. The key point to note is that, after resampling, the distribution represented by the particle set is unchanged. This is formalised in the next result, which is stated as a conjecture because I have not yet found a rigorous proof for it.

Conjecture 7.1 *Let S' be a particle set obtained from S by weighted resampling with respect to a strictly positive, continuous importance function g. If S represents a distribution $p(\mathbf{x})$, then so does S'.*

Persuasive argument. Take $S = (\mathbf{x}_i^{(n)}, \pi_i^{(n)})_{i=1}^n, n = 1, 2, \ldots$, as in the above definition, but for convenience fix a value of n and drop the superscript ns. Set $\rho_i = g(\mathbf{x}_i)/\sum_j g(\mathbf{x}_j)$, obtain the $\mathbf{x}_i', \pi_i'$ by weighted resampling according to the definition, and set $\tilde{\pi}_i = \pi_j/\rho_j$ — so the $\tilde{\pi}_i$ are the *unnormalised* versions of the final weights π_i'. Set $K = \sum_{i=1}^n \tilde{\pi}_i$. By Lemma 2.7, we know that as $n \to \infty$, $K/n \to 1$ weakly.

Define indices $k_1, k_2, \ldots$ so that $\mathbf{x}_i' = \mathbf{x}_{k_i}$. Then by the result just mentioned we know the *normalised* weight π_i' is approximately $\pi_{k_i}/n\rho_{k_i}$. To complete the "proof" of the conjecture then, it will be enough to show that the total weight assigned to a value s_i is the same (as $n \to \infty$) in the initial and final particle sets. But this is now immediate: there are approximately $n\rho_{k_i}$ values equal to bx_i', and each has final weight $\pi_{k_i}/n\rho_{k_i}$. Thus the total weight assigned to $\mathbf{x}_i'$ is $n\rho_{k_i} \times \pi_{k_i}/n\rho_{k_i} = \pi_{k_i}$, just as in the initial particle set S. ■

Fortunately this argument can be made rigorous in a special case which always holds for the problems we are interested in.

Theorem 7.2 *Let $S' = (\mathbf{x}_i^{(n)'}, \pi_i^{(n)'})_{i=1}^n$, $n = 1, 2, \ldots$, be a particle set obtained from $S = (\mathbf{x}_i^{(n)}, \pi_i^{(n)})_{i=1}^n$, $n = 1, 2, \ldots$, by weighted resampling with respect to a strictly positive, continuous importance function g. Suppose in addition that there is a continuous, strictly positive function $f(\mathbf{x})$ such that*

$$\pi_i^{(n)} = \frac{f(\mathbf{x}_i^{(n)})}{\sum_{j=1}^n f(\mathbf{x}_j^{(n)})}. \tag{7.3}$$

Then if S represents a distribution $p(\mathbf{x})$, so does S'.

Remark. The additional assumption of equation 7.3 holds, in particular, immediately after a standard resampling operation, since then all the weights are equal and we can take $f \equiv 1$. If some dynamics are then applied, the condition still holds since the particles still have equal weights. The condition also holds when an equally weighted particle set has been multiplied by a continuous positive function h, since then we just take $f = h$.

Proof. If is easy to check that when the condition of equation 7.3 holds, the Condensation diagram (7.2) is equivalent to

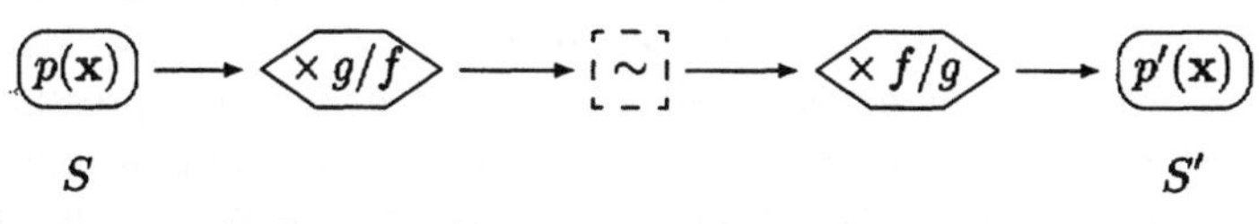

(7.4)

where the $\sim$ signifies standard resampling. By Theorem 2.4, the standard resampling operation does not alter the distribution represented. Hence the two multiplications in equation 7.4 cancel and the distribution represented is unchanged, so that $p' = p$. ■

7.3 Basic partitioned sampling

Let us return to the problem of tracking two targets, A and B. If each target deforms and moves in a space of d dimensions, there are $2d$ dimensions to be inferred at each time step. By employing partitioned sampling, this problem will be reduced to the more feasible task of performing two inferences of d dimensions each. To be more concrete, suppose it is known that target A partially occludes target B. Then we can localise the two targets efficiently by first inferring the configuration of target A, and then using this knowledge to localise B. To infer the configuration of A, we will use the one-body density $\mathcal{P}_A$ defined by equation 6.8 (page 118).

The basic algorithm is as follows. Suppose we can decompose the joint dynamics as

$$p(\mathbf{x}''|\mathbf{x}) = \int_{\mathbf{x}'} p_B(\mathbf{x}''|\mathbf{x}')p_A(\mathbf{x}'|\mathbf{x})d\mathbf{x}',$$

where p_A are the dynamics for target A and similarly for B. (This assumption would hold if, as is often the case, the dynamics of the targets were independent of each other.) Then one time step of the partitioned sampling algorithm is defined by the following diagram:

$$\underset{(a)}{\boxed{S^t}} \longrightarrow \boxed{\sim} \longrightarrow \underset{(b)}{\langle *p_A(\mathbf{x}'|\mathbf{x})\rangle} \longrightarrow \boxed{\sim \mathcal{P}_A} \longrightarrow \langle *p_B(\mathbf{x}''|\mathbf{x}')\rangle \longrightarrow \underset{(c)}{\langle \times \mathcal{P}(\mathbf{Z}^{t+1}|\mathbf{x}'')\rangle} \longrightarrow \underset{(d)}{\boxed{S^{t+1}}} \tag{7.5}$$

The symbol $\sim \mathcal{P}_A$ means "perform weighted resampling with respect to the importance function $\mathcal{P}_A$", and the labels (a)–(d) refer to the example given later in Figure 7.5. The validity of this algorithm is guaranteed by the following

Proposition 7.3 *If the dynamics $p(\mathbf{x}''|\mathbf{x})$ can be decomposed as $p(\mathbf{x}''|\mathbf{x}) = \int_{\mathbf{x}'} p_B(\mathbf{x}''|\mathbf{x}')p_A(\mathbf{x}'|\mathbf{x})\, d\mathbf{x}$, the posterior generated by the diagram of equation 7.5 is the same as that generated by the diagram of equation 6.7 (page 118).*

Proof. This Condensation diagram is well-formed, since every application of dynamics is immediately preceded by a resampling operation. (It is easy to check that weighted resampling can replace standard resampling in the condition for Theorem 2.3 to hold.) So by the Condensation theorem we know the operations shown have the expected effect on the distribution represented by the particle sets. Moreover, by Theorem 7.2, the weighted resampling operation $\sim \mathcal{P}_A$ has no effect on the distribution represented. Hence we may delete this step from the diagram without affecting the posterior. The step $*p_A(\mathbf{x}'|\mathbf{x})$ is now followed immediately by $*p_B(\mathbf{x}''|\mathbf{x}')$, and by assumption the consecutive application of these steps is equivalent to $*p(\mathbf{x}''|\mathbf{x})$. Making this substitution on the diagram, we obtain the diagram of equation 6.7, as desired. ∎

Remark. It is clear from the proof that instead of $\mathcal{P}_A$ in equation 7.5, one could use any strictly positive function without affecting the posterior. However, the objective of partitioned sampling is to obtain an accurate representation of the posterior with a moderate number of particles. Hence we would like the weighted resampling step to position as many particles as possible near peaks in the posterior. Because we assumed target A partially occludes target B, the one-body density $\mathcal{P}_A$ is a good choice as importance reweighting function. Particles surviving the weighted resampling step lie in peaks of $\mathcal{P}_A$, and this function has peaks in the "right" place because target A is completely visible.

Example Consider the simple example from Section 2.1, in which $\mathcal{X}$ is a two-dimensional configuration space; then each particle in a particle set can be represented on a plane, with area proportional to its weight. Figure 7.4 (which is a partial replication of Figure 2.2) uses this convention to illustrate one iteration of the conventional (non-partitioned) Condensation algorithm. Box (a) shows the prior — a Gaussian centred on the centre of the image. The black cross shows the actual position of the target, which of course is not known to the algorithm at this stage. Box (b) shows the distribution after the prior has been resampled and the dynamics (which in this case are isotropic additive Gaussian) have been applied. Note that at this point each particle has equal weight. In (c), the particles have the same configurations as in (b), but their weights are now proportional to the observation density. This is the particle representation of the posterior distribution.

Figure 7.5 shows the application of partitioned sampling in the same scenario. The dynamics and observations are partitioned into $\mathbf{x}^A$ and $\mathbf{x}^B$ components. Box (a) shows the same prior as in Figure 7.4. In (b), the prior has been resampled and the $\mathbf{x}^A$-component of the dynamics has been applied. To produce (c), we first perform weighted resampling on these particles, with respect to an importance function centred on an observation of the $\mathbf{x}^A$-coordinate of the target. Recall that this has no effect on the distribution represented, but of course it selects many particles whose $\mathbf{x}^A$-coordinate is close to the target's — this will be beneficial later. Next the $\mathbf{x}^B$-component of the dynamics is applied, producing the particle set shown in (c). Finally, this set is multiplied by the joint observation density for $\mathbf{x}^A$ and $\mathbf{x}^B$ coordinates. The result is shown in (d). Notice how dense this representation is, compared to the final outcome of non-partitioned sampling in Figure 7.4.

7.4 Branched partitioned sampling

Branching is a refinement of partitioned sampling which is needed in our application to a mixed state Condensation tracker. In the discussion above, it was assumed target A partially occluded target B. This enabled us to select the one-body density $\mathcal{P}_A$ as a suitable importance function for the weighted resampling step in equation 7.5. However, at any given time step, there are some particles for which $y = 1$ (i.e. A is unoccluded) and some for which $y = 2$ (i.e. B is unoccluded). It would be preferable to select a *different* importance

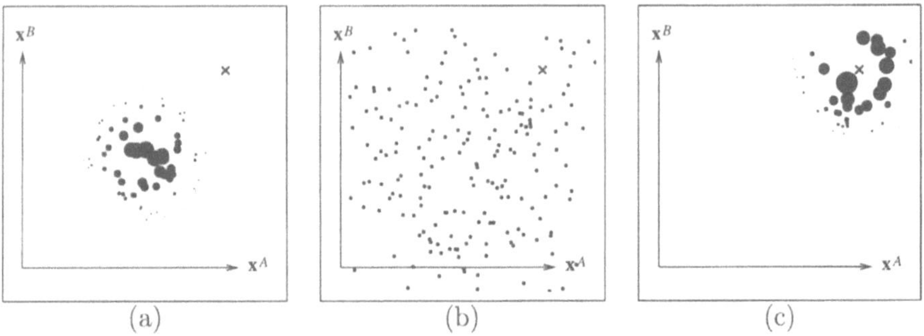

Figure 7.4: ***Conventional (i.e. non-partitioned) Condensation.*** *(See also Figure 2.2.) The true position of the target in this two-dimensional configuration space is shown as a cross; particles representing a probability distribution are shown as circles whose areas are proportional to their weights. Each step shown is one stage in the Condensation diagram of equation 6.7. (a) The prior. (b) After the dynamics have been applied. (c) After reweighting by the posterior. The posterior is centred at approximately the correct position, but this representation of the posterior is not very accurate because relatively few particles have significant weights. In technical terms, the survival diagnostic given by equation 7.7 is low. Superior results are achieved using partitioned sampling (Figures 7.5 and 7.6).*

function for each y value.

This is achieved by the *branched* partitioned sampling algorithm summarised on the following diagram:

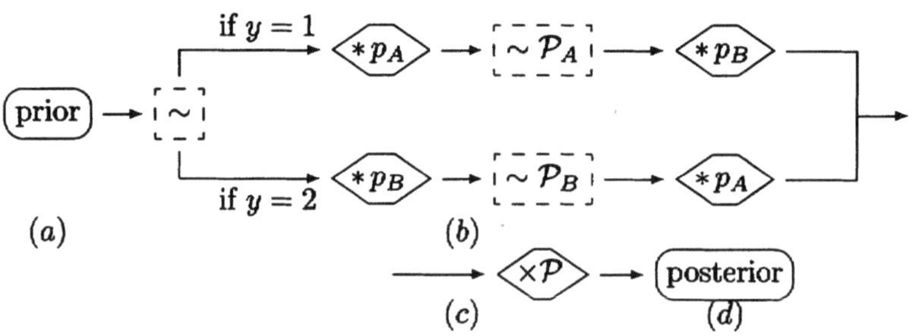

(7.6)

Particles for which $y = 1$ follow the top path, which positions the $\mathbf{x}^A$-components first (near peaks in $\mathcal{P}_A$), since these particles believe A is unoccluded. Particles for which $y = 2$ follow the bottom path, since they believe B is unoccluded. The final result is that many more particles survive the resampling process, compared to the non-partitioned process, and the posterior is represented more accurately.

One technical point: the sum of weights π_i in any one branch need not be unity. Hence when performing weighted resampling, the new weights must be normalised to have the same sum as before the resampling.

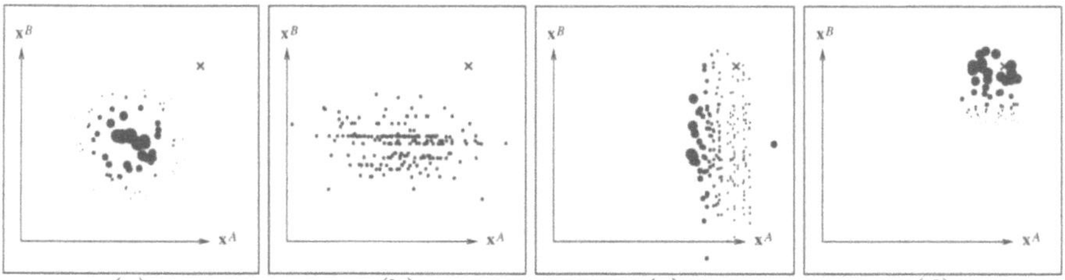

Figure 7.5: ***Partitioned sampling.*** *A simple example implementing the Condensation diagram of equation 7.5. The two-dimensional configuration space is partitioned as the cross product of the* $\mathbf{x}^A$ *and* $\mathbf{x}^B$ *dimensions, and the true position of the target is shown as a cross. (a) The prior. (b) Dynamics have been applied in the* $\mathbf{x}^A$*-direction. (c) The weighted resampling operation has been performed, and the remaining dynamics applied. (d) The particles are reweighted by the posterior. Note how fine-grained the sample set for the posterior is, compared with the final set from conventional sampling in Figure 7.4. In other words, this representation of the posterior has a higher survival diagnostic (equation 7.7) than that in Figure 7.4.*

Example In Figure 7.6, the two-dimensional example has been augmented to include a binary discrete label, indicated by the colour of each particle (grey or black). The prior, (*a*), gives an equal weighting to the two discrete states. Box (b) shows the particle set immediately after the branching: grey particles have had the $\mathbf{x}^B$-component of the dynamics applied to them, whereas black particles have received the $\mathbf{x}^A$-component. Box (c) shows the particle set after the branches merge again. The grey particles receive weighted resampling with respect to an observation of the target's $\mathbf{x}^B$-coordinate, while the black particles receive weighted resampling with respect to an observation of the target's $\mathbf{x}^A$-coordinate. Then the remaining dynamics are applied: the $\mathbf{x}^A$ component to the grey particles, and the $\mathbf{x}^B$ component to the black particles. This results in (c). Finally, the weights are multiplied by the joint observation density for $\mathbf{x}^A$ and $\mathbf{x}^B$, producing the posterior shown in (d).

7.5 Performance of partitioned sampling

Evaluating the performance of particle filters such as Condensation is a difficult problem (Carpenter et al., 1997; Doucet, 1998; Liu and Chen, 1998). To compare the two schemes given in equations 7.6 and 6.7 we use the survival diagnostic $\mathcal{D}(n)$ introduced in Section 2.6.1. It is defined for a set of particles with weights $\pi_1, \ldots \pi_n$ as

$$\mathcal{D}(n) = \left(\sum_{i=1}^{n} \pi_i^2 \right)^{-1}. \tag{7.7}$$

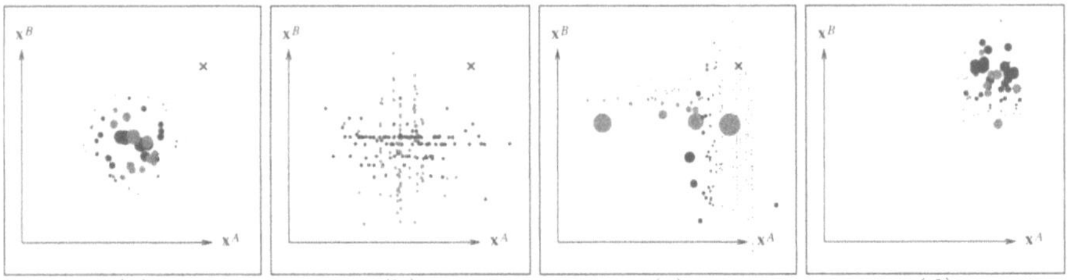

Figure 7.6: ***Branched partitioned sampling.*** *Each step shows a stage from the Condensation diagram of equation 7.6. The two-dimensional configuration has been augmented with a binary variable y, shown as black ($y = 1$) or grey ($y = 2$), and the value of this variable determines which branch is taken in equation 7.6. (a) The prior. (b) Dynamics have been applied in the* $\mathbf{x}^A$*-direction for black particles and the* $\mathbf{x}^B$*-direction for grey particles. (c) The weighted resampling operation has been performed, and the remaining dynamics applied. (d) The particles are reweighted by the posterior. The effective sample size of the posterior is greater than for the unpartitioned method (Figure 7.4) but, in this simple example, is no better than the non-branched, partitioned method (Figure 7.5). However, that is because this example is symmetric in A and B: the branched method* would *be superior if the two importance functions* $\mathcal{P}_A, \mathcal{P}_B$ *used to produce (c) were not equally good predictors of particle position.*

Intuitively, this corresponds to the number of "useful" particles: if all have the same weight $1/n$ then $\mathcal{D} = n$, whereas if all but one of the weights are negligible we have $\mathcal{D} = 1$. Any other distribution of the weights falls between these two extremes. Figure 7.7 compares $\mathcal{D}(n)$ for the conventional ("unpartitioned") and partitioned methods. It is clear that partitioned sampling achieves much higher values of $\mathcal{D}$ than unpartitioned sampling and that we can therefore expect much better tracking performance for the same computational expense. We can show this is indeed the case in a practical example: Figure 7.8 shows stills from a certain sequence tracked by each method. With partitioned sampling, and $n = 750$ particles, the tracking succeeds. However, despite using four times as many particles, unpartitioned sampling fails to track on the same sequence.

7.6 Partitioned sampling for articulated objects

Partitioned sampling is not restricted to improving the efficiency of multiple object tracking. In fact, it can be used whenever the following conditions hold.

- The configuration space $\mathcal{X}$ can be partitioned as a Cartesian product $\mathcal{X} = \mathcal{X}_1 \times \mathcal{X}_2$.
- The dynamics d can be decomposed as $d = d_1 * d_2$, where d_2 acts on $\mathcal{X}_2$. This means that if $\mathbf{x} = (\mathbf{x}_1, \mathbf{x}_2)$ and $\mathbf{x}' = (\mathbf{x}'_1, \mathbf{x}'_2)$ with $\mathbf{x}_i, \mathbf{x}'_i \in \mathcal{X}_i$, and

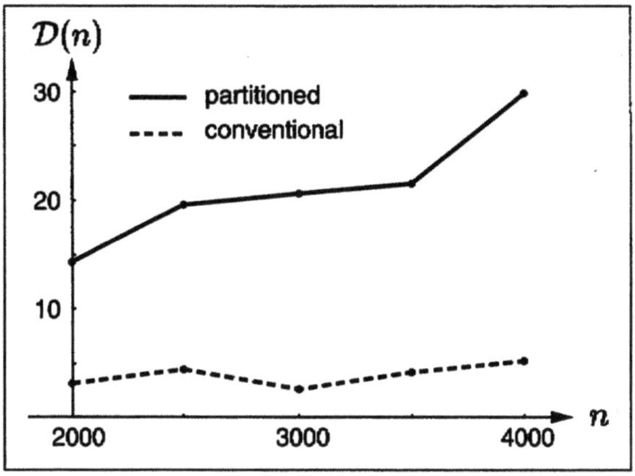

Figure 7.7: ***Survival diagnostic $\mathcal{D}(n)$ for partitioned and conventional (unpartitioned) sampling methods.*** *The graph shows the value of $\mathcal{D}(n)$ following a ten-frame sequence tracking two leaves, averaged over five runs. Note the superior performance of the partitioned sampling method.*

$\mathbf{x}'$ is a random draw from $d_2(\cdot|\mathbf{x})$, then $\mathbf{x}'_1 = \mathbf{x}_1$. Informally, the second partition of the dynamics does not change the value of the projection of any particle into the first partition of the configuration space.

- An importance function g_1 defined on $\mathcal{X}_1$ is available, which is peaked in the same region as the posterior restricted to $\mathcal{X}_1$.

There is also an obvious generalisation to $k > 2$ partitions: the configuration space is partitioned as $\mathcal{X} = \mathcal{X}_1 \times \ldots \times \mathcal{X}_k$, the dynamics as $d = d_1 * \ldots * d_k$ with each d_j acting on $\mathcal{X}_j \times \ldots \times \mathcal{X}_k$, and importance functions $g_1, g_2, \ldots g_{k-1}$ with each g_j peaked in the same region as the posterior restricted to $\mathcal{X}_j$.

One example of such a system is an articulated object. The example given in this section is of a hand tracker which models the fist, index finger and thumb as an articulated rigid object with three joints. This system was implemented in collaboration with Michael Isard (MacCormick and Isard, 2000; Isard and MacCormick, 2000).

The partitioned sampling algorithm used for this application is shown in the following Condensation diagram:

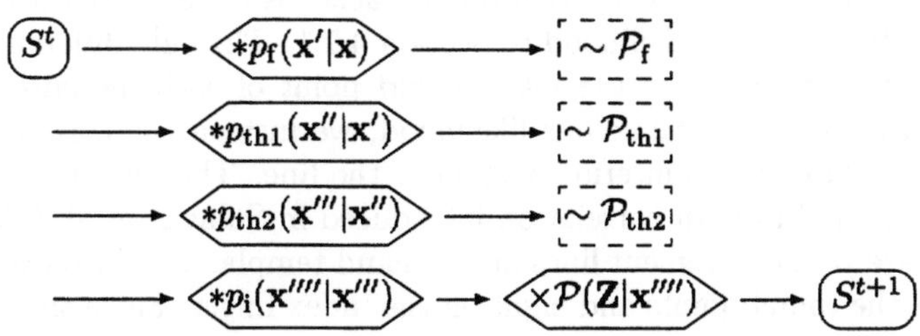

(7.8)

partitioned, $n = 750$

unpartitioned, $n = 3000$

Figure 7.8: ***Unpartitioned sampling can fail when partitioned sampling does not, even if more particles are used.*** *The final frame from a tracked sequence is shown: with unpartitioned sampling, the tracking fails despite using four times as many particles as the partitioned method.*

The subscript 'f' stands for "fist", 'th1' for "first thumb joint", 'th2' for "second thumb joint", and 'i' for "index finger". So the configuration space is partitioned into four parts:

- $\mathcal{X}_{\mathrm{f}} \equiv$ scale, orientation, and x and y translation of the fist
- $\mathcal{X}_{\mathrm{th1}} \equiv$ joint angle of base of thumb
- $\mathcal{X}_{\mathrm{th2}} \equiv$ joint angle of tip of thumb
- $\mathcal{X}_{\mathrm{i}} \equiv$ joint angle of index finger.

The dynamics are decomposed as $d = d_{\mathrm{f}} * d_{\mathrm{th1}} * d_{\mathrm{th2}} * d_{\mathrm{i}}$ with the last three operations consisting of a deterministic shift plus Gaussian diffusion within the appropriate partition only. Note that although $\mathcal{X}$ is a shape space of splines, it is not given by the linear parameterisation (equation 1.2) normally used for shape spaces. Instead it is parameterised by the seven physical variables listed above (scale, orientation, x and y translation, and the three joint angles), so that any $\mathbf{x}$ is an element of $\mathbb{R}^7$.

Likelihood function and importance functions

The likelihood $\mathcal{P}(\mathbf{Z}|\mathbf{x})$ used by this hand tracker is closely related to the robustified colour contour likelihood of Section 4.7.1. The only difference is that instead of a single pixel at the interior end point of each measurement line being used as input to the colour likelihood, we instead use correlation with a colour template on the interior portion of the line. This can be viewed as a precursor to the Bayesian correlation introduced in Sullivan et al. (1999).

There are 28 measurement lines on the hand template: eight on the fist, six on each of the thumb joints and eight on the index finger. Since the likelihood factorises as a product of likelihoods for individual measurement lines, this gives us a convenient way to re-express the likelihood:

$$\mathcal{P}(\mathbf{Z}|\mathbf{x}) = \mathcal{P}_{\mathrm{f}}(\mathbf{Z}_{\mathrm{f}}|\mathbf{x}_{\mathrm{f}})\,\mathcal{P}_{\mathrm{th1}}(\mathbf{Z}_{\mathrm{th1}}|\mathbf{x}_{\mathrm{f}},\mathbf{x}_{\mathrm{th1}})\,\mathcal{P}_{\mathrm{th2}}(\mathbf{Z}_{\mathrm{th2}}|\mathbf{x}_{\mathrm{f}},\mathbf{x}_{\mathrm{th2}})\,\mathcal{P}_{\mathrm{i}}(\mathbf{Z}_{\mathrm{i}}|\mathbf{x}_{\mathrm{f}},\mathbf{x}_{\mathrm{i}}), \quad (7.9)$$

where, for example, $\mathbf{Z}_{\mathrm{f}}$ are the measurements on the eight fist locations, $\mathbf{x}_{\mathrm{f}}$ are the components of $\mathbf{x}$ which specify the configuration of the fist, and similarly for the other subscripts.

The factorisation of equation 7.9 immediately suggests the use of $\mathcal{P}_{\mathrm{f}}$, $\mathcal{P}_{\mathrm{th1}}$ and $\mathcal{P}_{\mathrm{th2}}$ as importance functions, since they should be peaked at the correct locations of the fist and thumb joints one and two respectively. This is precisely what the implementation does; hence the presence of $\mathcal{P}_{\mathrm{f}}$, $\mathcal{P}_{\mathrm{th1}}$ and $\mathcal{P}_{\mathrm{th2}}$ on equation 7.8.

As a final point, note that in the particular case in which the overall likelihood $\mathcal{P}$ can be expressed as a product of the importance functions and another easily calculated function (in this case, $\mathcal{P}_{\mathrm{i}}$), the diagram in equation 7.8 can be given an even simpler form which uses standard resampling rather than weighted resampling:

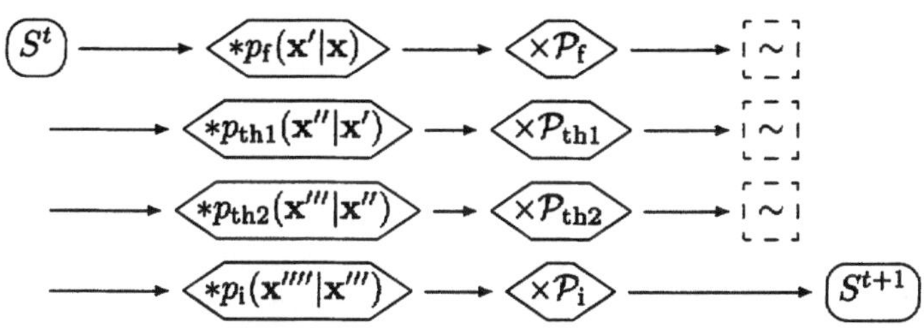

(7.10)

Dividing effort between the partitions

An important advantage of partitioned sampling is that the number of particles devoted to each partition can be varied. Partitions which require a large number of particles for acceptable performance can be satisfied without incurring additional effort in the other partitions. For instance, in the hand tracking application the fist often moves rapidly and unpredictably, whereas the joint angles of finger and thumb tend to change more slowly. Hence we use $n_1 = 700$ particles for the fist partition, but only $n_2 = n_3 = 100$ particles for the two thumb partitions and $n_4 = 90$ for the index finger partition. A glance at the diagram in equation 7.10 shows this produces a substantial saving, since at every time-step we avoid calculating $\mathcal{P}_{\mathrm{th1}}(\mathbf{Z}_{\mathrm{th1}}|\mathbf{x})$, $\mathcal{P}_{\mathrm{th2}}(\mathbf{Z}_{\mathrm{th2}}|\mathbf{x})$ and $\mathcal{P}_{\mathrm{i}}(\mathbf{Z}_{\mathrm{i}}|\mathbf{x})$ for over 600 values of $\mathbf{x}$ that would otherwise have been required.

Other details

Initialisation and re-initialisation are handled by the ICondensation mechanism of Isard and Blake (1998b). Various standard tools, such as background subtraction (which can be performed on an SGI Octane very cheaply using the alpha-blending hardware), and least-squares fitting of an auxiliary spline to the tip of the index finger, are used to refine the performance of the tracker.

Figure 7.9: ***Heavy clutter does not hinder the Condensation tracker.*** *Even moving the papers on the desk to invalidate the background subtraction does not prevent the Condensation tracker functioning. The fingertip localisation is less robust, however, and jitter increases in heavily cluttered areas.*

Our objective here is to present a practical example of partitioned sampling for articulated objects, so these additional details are relegated to a technical report (Isard and MacCormick, 2000).

7.6.1 Results: a vision-based drawing package

The hand tracker described in the previous section was implemented on an SGI Octane with a single 175 MHz R10000 CPU. Using 700 samples for the hand base, 100 samples for each of the thumb joints and 90 samples for the index finger, the tracker consumes approximately 75% of the machine cycles, which allows real-time operation at 25 Hz with no dropped video frames even while other applications are running on the machine. The tracker is robust to clutter (Figure 7.9), including skin-coloured objects (Figure 7.10). The position of the index finger is located with considerable precision (Figure 7.11) and the two articulations in the thumb are also recovered with reasonable accuracy (Figure 7.12).

We have developed a simple drawing package to explore the utility of a

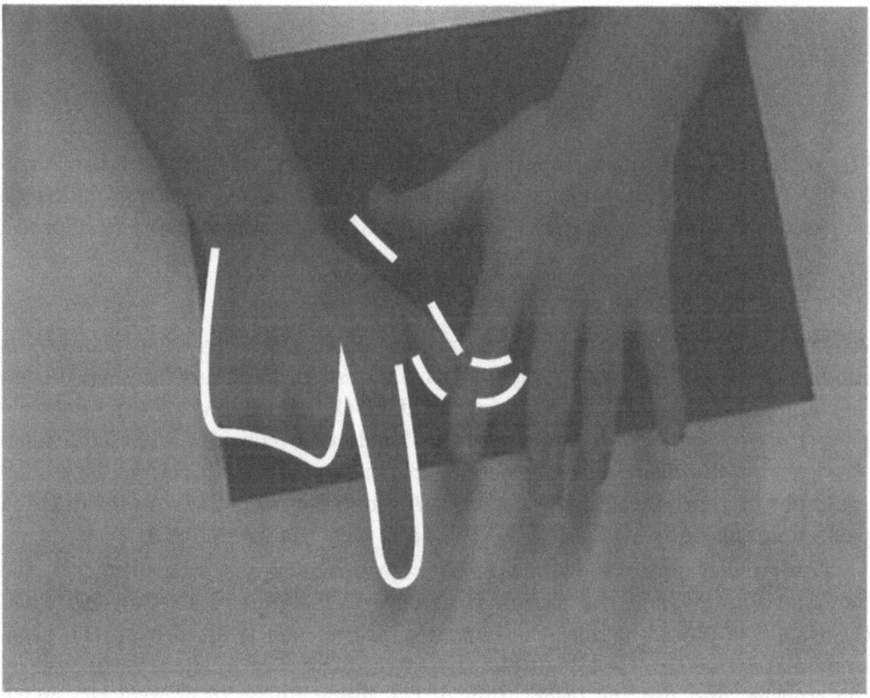

Figure 7.10: ***Skin-coloured objects do not distract the tracker.*** *Here two hands are present in the image, but tracking remains fixed to the right hand. If the right hand were to leave the field of view the tracker would immediately reinitialise on the left hand.*

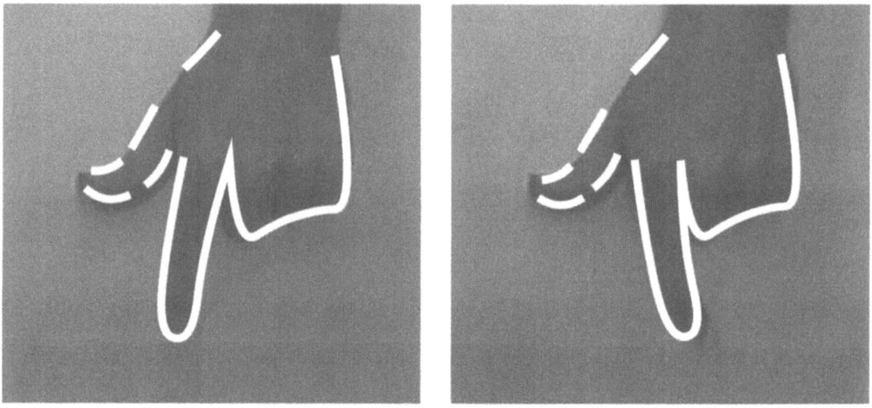

Figure 7.11: ***The index finger is tracked rotating relative to the hand body.*** *The angle of the finger is estimated with considerable precision, and agile motions of the fingertip, such as scribbling gestures, can be accurately recorded.*

vision-based hand tracker for user-interface tasks. The tracking achieved is sufficiently good that it can compete with a mouse for freehand drawing, though (currently) at the cost of absorbing most of the processing of a moderately powerful workstation. It is therefore instructive to consider what additional strengths of the vision system we can exploit to provide functionality which could not be reproduced using a mouse. A video sequence showing the drawing package in action is available from the book web site (see page 3).

The current prototype drawing package provides only one primitive, the freehand line. When the thumb is extended, the pointer draws, and when the thumb is placed against the hand the virtual pen is lifted from the page. Immediately we can exploit one of the extra degrees of freedom estimated by the tracker, and use the orientation of the index finger to control the width of the line being produced. When the finger points upwards on the image, the pen draws with a default width; then, as the finger rotates, the width varies from thinner (finger anticlockwise) to thicker (finger clockwise) — see Figure 7.13. The scarcity of variable-thickness lines in computer-generated artwork is a testament to the difficulty of producing this effect with a mouse.

The fact that a camera is observing the desk also allows other intriguing features not directly related to hand tracking. We have implemented a natural interface to translate and rotate the virtual workspace for the modest hardware investment of a piece of black paper (Figure 7.14). The very strong white-to-black edges from the desk to the paper allow the paper to be tracked with great precision using a simple Kalman filter, at low computational cost. Translations and rotations of the paper are then reflected in the virtual workspace, a very satisfying interface paradigm. While one hand draws, the other hand can adjust the workspace to the most comfortable position. In the future it should be possible to perform discrete operations, such as switching between drawing

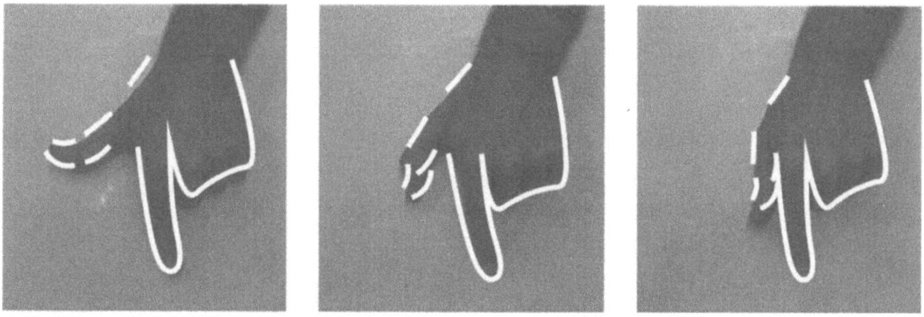

Figure 7.12: ***The two degrees of freedom of the thumb are tracked.*** *The thumb angles are not very reliably estimated. This is probably partly because the joints are short, and so offer few edges to detect, and more importantly because the shape model gives a poor approximation to the thumb when it opposes. The gross position of the thumb can be extracted consistently enough to provide a stable switch which can be used analogously to a mouse button.*

tools, using simple static gesture recognition on one of the hands. Tracking both hands would allow more complex selection tasks, for example continuous zooming, or colour picking.

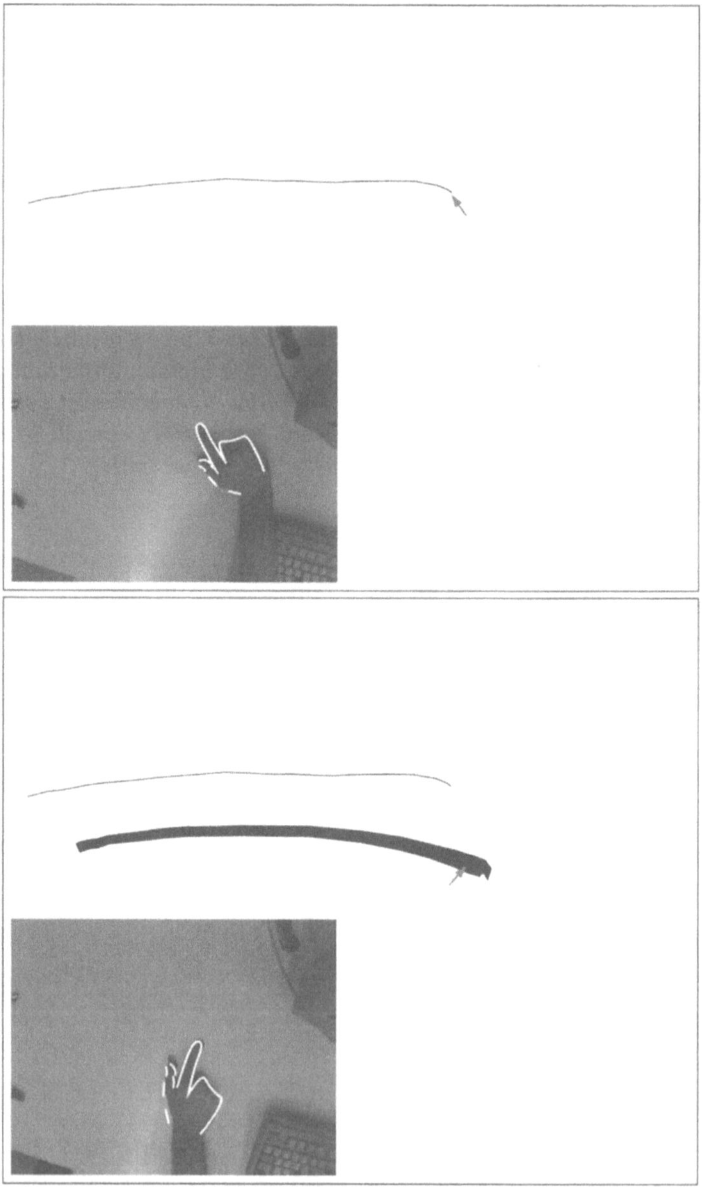

Figure 7.13: ***Line thickness is controlled using the orientation of the index finger.*** *The top image shows a line drawn with the index finger pointing to the left, producing a thin trace. In the bottom image the finger pointed to the right and the line is fatter. Of course, if the finger angle varies while the line is being drawn, a continuous variation of thickness is produced.*

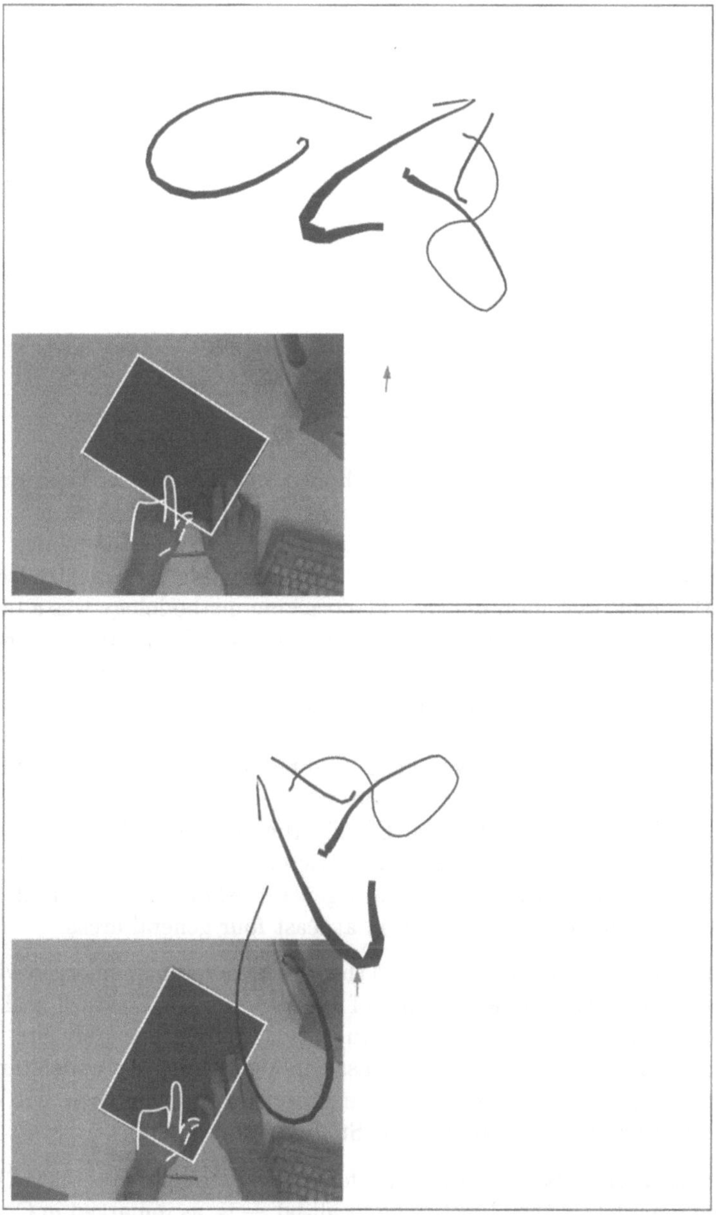

Figure 7.14: ***Moving around the virtual workspace is accomplished by following the tracked outline of a physical object.*** *The piece of black paper can be tracked with a simple Kalman filter, and the virtual drawing follows the translations and rotations of the paper. The virtual workspace has been rotated anticlockwise between the top and bottom frames. This is a very natural interface which can be used while drawing.*

8

Conclusion?

Readers looking for a summary of the book can do no better than turn back to Chapter 1, where I hope they will get to "know the place for the first time" (Eliot, 1942). The book essentially outlined a set of models termed contour likelihoods, which can be used for performing statistical inference in localisation and tracking problems. The tracking methodology was a technique called particle filtering. The basic theory of particle filters was presented, together with some methods of analysing the performance of particle filters, and a technique termed partitioned sampling which can dramatically improve the performance of particle filters in an important class of tracking problems.

The theme throughout was the use of probabilistic modelling and stochastic algorithms, so it is interesting to wonder what sort of future this approach will have in the field of computer vision. Clearly, the range of problems which can be solved by the current methods is very restricted (compared to the performance of humans), so one possible answer is a general exhortation to "do it better". The improvements can be expected in at least four general areas.

Background models Mathematical theories have recently emerged which appear to capture at least some of the generic properties of real images (Mumford and Gidas, 1999; Wu et al., 1999), but they are not yet tractable enough for use in localisation algorithms. As understanding of these models increases, perhaps methods for applying them will emerge; a recent step in this direction is Sullivan et al. (1999).

Foreground models The problem of foreground modelling — that is, specifying a model for the object to be localised — is, as remarked in Chapter 1, ultimately linked with much more general AI problems of language and consciousness. However, even if we ignore this, there is much work to be done in unifying the many types of foreground model into an ever more rigorous probabilistic framework. In particular, one might hope for a unified, probabilistic theory of texture, shape, and correlation.

Learning Learning algorithms have played almost no part in this book (although much of the tracking relies on learning techniques that have been published elsewhere (e.g. Blake and Isard, 1998) for the dynamics and shape spaces). However, many other approaches to localisation and tracking do employ learning extensively — some with a more sound statistical underpinning than others, of course. An ongoing challenge will be to extend the models of this book in such a way that statistical rigour is preserved while allowing the increased flexibility permitted by learnt models.

Particle filters Many open problems remain in the domain of particle filters. The design of partitioned sampling schemes is very much an art at the moment: can we develop a theory which elucidates the choices behind partitioned sampling? In particular, how should particles be allocated between partitions to make the most out of a fixed computational resource? Is it possible to design partitioned sampling schemes on tree structures which do not needlessly impoverish some of the branches? More generally, there are some ideas from the MCMC literature which may be applicable to particle filters: the annealed importance sampling of Neal (1998) is one example.

In any case, it seems likely that probability and statistics will continue to be indispensable tools, as we edge closer to designing algorithms that can make computers see.

A

Appendix

A.1 Measures and metrics on the configuration space

There were several occasions in this book when a natural measure on the configuration space $\mathcal{X}$ was needed. The first was in Section 4.4, where a default uniform prior was required. We explain below why a particular measure called the Haar measure should be used for this.

The Haar measure. This can be defined for any $\mathcal{X}$ that is a sub-manifold of some locally compact topological group $\mathcal{W}$. (This is true of all contour configuration spaces.) A theorem from measure theory tells us that for every such $\mathcal{W}$ there exists a unique Borel measure μ, called the *Haar measure*, with the property that for all Borel sets $S \subset \mathcal{W}$ and all $w \in \mathcal{W}$, $\mu(E) = \mu(wE)$. Here wE denotes the set $\{ws : s \in S\}$, i.e. the image of S under left multiplication by w. Of course the uniqueness is up to a scale factor. A moment's thought reveals the above condition coincides precisely with our intuitive notion of a uniform density. Although no general method for constructing Haar measures is known, the construction is obvious for all contour configuration spaces. For example, the familiar uniform distribution on the groups of translations and 2D rotations are Haar measures. For the group of 3D rotations, use the well-known isomorphism to the group of (pairs of) quaternions of modulus 1. Now regard this manifold as a 3-sphere embedded in $\mathbb{R}^4$; it inherits a measure from the Euclidean metric on $\mathbb{R}^4$ which is easily shown to be the Haar measure.

The spline metric and spline measure. We define the *spline metric* $d(\cdot,\cdot)$ on $\mathcal{X}$ following Blake et al. (1993). For reasons we need not discuss, it was called the Mahalanobis distance there. We set

$$d(\mathbf{x}_1, \mathbf{x}_2) = \int_s \|\mathbf{b}_1(s) - \mathbf{b}_2(s)\|^2 ds,$$

where $\mathbf{b}_i$ is the B-spline corresponding to $\mathbf{x}_i$ via equations 1.1 and 1.2, s is the B-spline parameter, and $\|\cdot\|$ denotes Euclidean distance in the image.

This choice of metric is justifiable on theoretical grounds, agrees very well with human assessment of curves being "nearly the same",[1] and can be calculated very rapidly.[2]

There is a canonical way to define a measure on $\mathcal{X}$ from this metric. (A metric need not induce a measure on a general topological space, but in this case everything works out nicely.) $\mathcal{X}$ is embedded in the finite-dimensional vector space V of splines with the given number of control points. The metric $d(\cdot,\cdot)$ above is induced in V by an easily-defined inner product (see Blake et al. (1993)). But an inner product on a finite dimensional vector space gives us the Lebesgue measure, and it is this measure which $\mathcal{X}$ can inherit from V.

A.2 Proof of the interior-exterior likelihood

To prove Proposition 3.4, we refer to the interior-exterior generative model explicitly described on page 47. Statement (a) is proved first. This means we can assume d is set to `True` in step 2. The proof consists of simply writing down the pdf of the model after each step.

- After step 1: $p(a) = \mathcal{G}(a|\nu)$.
- After step 2: For fixed d, we still have $p(a) = \mathcal{G}(a|\nu)$ as before.
- After step 3:

$$p(m; a, \phi_1, \dots \phi_m) = \begin{cases} \frac{f_a(m)\mathcal{G}(a|\nu)}{a^m} & \text{if } 0 \le \phi_1, \dots \phi_m \le a, \\ 0 & \text{otherwise.} \end{cases}$$

  The factor $1/a^m$ is just a normalising constant due to the fact that the features $(\phi_1, \dots \phi_m)$ are uniformly distributed over the hypercube $[0, a]^m$ of volume a^m.

- After step 4:

$$p(m, n; a, \phi_1, \dots \phi_m, \beta_1, \dots \beta_n) = \begin{cases} \frac{f_a(m) b_{L-a}(n) \mathcal{G}(a|\nu)}{(L-a)^n a^m} & \text{if } 0 \le \phi_1, \dots \phi_m \le a \le \beta_1, \dots \beta_n, \\ 0 & \text{otherwise.} \end{cases}$$

  The new factor $1/(L-a)^n$ is just another normalising constant reflecting the fact that the features $(\beta_1, \dots \beta_n)$ are uniformly distributed over the hypercube $[a, L]^n$.

---

[1] It is well known that curves which appear identical to a human can nevertheless be far apart according to the chosen metric. But this causes no problem for us since the reverse implication, which we do require, is always true: if the metric says two curves are nearly the same, a human will agree.

[2] Exploiting the fact that our curves are B-splines means the metric can be calculated using four matrix × vector multiplications, where the matrix is a precalculated, sparse square matrix whose order is the number of control points in the B-spline.

- After step 5: The basic idea to use here is that there are $N!$ equally likely permutations to be applied, but only $n!m!$ of them maintain the ordering restriction on the arguments. For convenience, write

$$\alpha(m,n,a) = \frac{f_a(m) b_{L-a}(n) \mathcal{G}(a|\nu)}{(L-a)^n a^m}, \tag{A.1}$$

and rename $(\phi_1, \ldots \phi_m, a, \beta_1, \ldots \beta_n)$ as $(y_1, \ldots y_N)$. Then the pdf after step 4 can be rewritten as

$$p(m,n;y_1,\ldots y_N) = \begin{cases} \alpha(m,n,y_{m+1}) & \text{if } 0 \le y_1, \ldots y_m \le \\ & y_{m+1} \le y_{m+2}, \ldots y_N, \\ 0 & \text{otherwise.} \end{cases}$$

Fix a permutation $\sigma = (\sigma_1, \ldots \sigma_N)$ of the numbers $(1, \ldots N)$; then writing $(z_1, \ldots z_N) = (y_{\sigma_1}, \ldots y_{\sigma_N})$ we have

$$p(m,n;z_1,\ldots z_N|\sigma) = \begin{cases} \alpha(m,n,z_{\sigma_{m+1}}) & \text{if } 0 \le z_{\sigma_1}, \ldots z_{\sigma_m} \le \\ & z_{\sigma_{m+1}} \le z_{\sigma_{m+2}}, \ldots z_{\sigma_N}, \\ 0 & \text{otherwise.} \end{cases}$$

According to step 5, each of the $N!$ possible permutations σ is equally likely. Define a subset S of permutations by the property that $\sigma \in S$ if $0 \le y_{\sigma_1}, \ldots y_{\sigma_m} \le y_{\sigma_{m+1}} \le y_{\sigma_{m+2}}, \ldots y_{\sigma_N}$. Then we have the pdf after step 5:

$$\begin{aligned} p(m,n;z_1,\ldots z_N) &= \frac{1}{N!} \sum_{\sigma} p(m,n;z_{\sigma_1},\ldots z_{\sigma_N}|\sigma) \\ &= \frac{1}{N!} \sum_{\sigma \in S} \alpha(m,n,z_{\sigma_{m+1}}) \\ &= \frac{m!n!}{N!} \alpha(m,n,z_{i_{m+1}}), \end{aligned} \tag{A.2}$$

where the last line follows because there are precisely $n!m!$ elements of S, and they all have $z_{\sigma_{m+1}} = z_{i_{m+1}} :=$ $(m+1)$th-smallest of the y_i.

- After step 6: summing equation A.2 over m and recognising that

$$\frac{m!n!}{N!} = \left(\binom{N}{m} (N-m) \right)^{-1}$$

gives the result.

To prove (b), we once again calculate the pdf after each step in the generative model on page 47. The first four steps are as in (a), then:

- After step 5: Take α as in equation A.1, but now rename $(\phi_1, \ldots \phi_m, \beta_1, \ldots \beta_n)$ as $(y_1, \ldots y_N)$. We can calculate a new pdf, conditional on a, as:

$$p(m,n;y_1,\ldots y_N|a) = \begin{cases} \alpha(m,n,a)/\mathcal{G}(a|\nu) & \text{if } 0 \le y_1, \ldots y_m \le \\ & a \le y_{m+1}, \ldots y_N, \\ 0 & \text{otherwise,} \end{cases}$$

and integrating out the hidden variable gives

$$p(m, n; y_1, \dots y_N) = \begin{cases} \int_{y_m}^{y_{m+1}} \alpha(m, n, a)\, da & \text{if } 0 \leq y_1, \dots y_m \leq \\ & y_{m+1}, \dots y_N, \\ 0 & \text{otherwise.} \end{cases}$$

As in part (a), there are $N!$ equally likely permutations to be applied, but only $n!m!$ of them preserve the ordering condition, meaning that the pdf of the permuted arguments after step 5 is

$$p(m, n; z_1, \dots z_N) = \frac{m!n!}{N!} \int_{z_{i_m}}^{z_{i_{m+1}}} \alpha(m, n, a). \qquad \text{(A.3)}$$

- After step 6: sum equation A.3 over m and we are done. It is easy to check the end cases, and to see that defining $z_{i_0} := 0$ and $z_{i_{N+1}} := L$ gives the correct result.

■

A.3 Del Moral's resampling lemma and its consequences

In this important appendix we collect together the proofs on particle sets which depend on the resampling concept. As will be seen, each follows from a technical lemma of Del Moral (1998) concerning a generalisation of the resampling concept in the space $\mathcal{P}(\mathcal{X})$. (See page 14 for the definition of $\mathcal{P}(\mathcal{X})$ and other notation used here.)

Let r_n denote the transition kernel on $\mathcal{P}(\mathcal{X})$ which sends a measure μ to $\frac{1}{n}\sum_{i=1}^{n} \delta_{\mathbf{x}_i}$, where the $\mathbf{x}_i$ are drawn from μ independently. So r_n is none other than the "resample n times" transition kernel on $\mathcal{P}(\mathcal{X})$ (alternatively, think of it as a function $r_n : \mathcal{P}(\mathcal{X}) \to \mathcal{P}(\mathcal{P}(\mathcal{X}))$). Given a bounded real-valued function F on $\mathcal{P}(\mathcal{X})$, define another function F_{r_n} on $\mathcal{P}(\mathcal{X})$ by setting

$$F_{r_n}(\nu) = \int F(\mu) r_n(\nu)(d\mu). \qquad \text{(A.4)}$$

We can now restate the important lemma from Del Moral (1998), which says (roughly speaking) that r_n is approximately the identity operation on $\mathcal{P}(\mathcal{X})$, for sufficiently large n. This will in turn prove to be exactly the property we need to discuss the effect of resampling in the space $\mathcal{P}(\mathcal{P}(\mathcal{X}))$.

Lemma A.1 *For every bounded real-valued function F on $\mathcal{P}(\mathcal{X})$, we have*

$$\lim_{n \to \infty} \|F_{r_n} - F\| = 0,$$

where $\| \bullet \|$ is the uniform norm.

The proof, which can be found in Del Moral (1998), is not overly technical but requires some additional concepts which would take us too far afield to reproduce here. Actually it does not require any deep measure-theoretic notions: the essential idea, familiar to all students of measure theory, is to prove the lemma for a class of simple F's whose linear combinations are dense in $\mathcal{P}(\mathcal{X})$.

The way is now clear to prove that random resampling on a particle set has no effect on the distribution represented.

Theorem A.2 *(This is the same as Theorem 2.4, page 21.) Let S be a particle set, and let S' be a particle set obtained from S by random resampling. If S represents a distribution $p(\mathbf{x})$, then so does S'.*

Proof. Recall r_n is the "resampling" transition kernel on $\mathcal{P}(\mathcal{X})$ defined above. Its more abstract uncle R_n is an operator on the $\Phi \in \mathcal{P}(\mathcal{P}(\mathcal{X}))$ defined as follows:

$$\begin{aligned} R_n : \mathcal{P}(\mathcal{P}(\mathcal{X})) &\to \mathcal{P}(\mathcal{P}(\mathcal{X})) \\ \Phi &\mapsto \int r_n(\mu)\Phi(d\mu). \end{aligned} \tag{A.5}$$

One can check this agrees with the definition of resampling a particle set, so the nth element of S' is given by $S'_n = R_n S_n$. To prove the theorem we have to show that for every F as above,

$$\lim_{n\to\infty} R_n S_n F - \mathbf{p}F = 0,$$

or in other words that

$$\int F(r_n\mu)S_n(d\mu) \to F(p).$$

The proof now proceeds with a standard "add and subtract" argument:

$$\int F(r_n\mu)S_n(d\mu) = \left(\int F(r_n\mu)S_n(d\mu) - \int F(\mu)S_n(d\mu)\right) + \int F(\mu)S_n(d\mu).$$

The first bracketed term here tends to zero by Lemma A.1, and the second term tends to $F(p)$ by the definition of a particle set. ■

The lemma can also be used to prove the basic result on standard Monte Carlo particle sets.

Corollary A.3 *(This is the same as Proposition 2.1, page 16.) A distribution $p(\mathbf{x})$ is represented by its standard Monte Carlo particle set.*

Proof. As usual, let $\mathbf{p}$ denote the element of $\mathcal{P}(\mathcal{P}(\mathcal{X}))$ which assigns mass one to $p(\mathbf{x}) \in \mathcal{P}(\mathcal{X})$ and zero to all other distributions. Recall the definitions of the resampling transition kernel $r_n : \mathcal{P}(\mathcal{X}) \to \mathcal{P}(\mathcal{P}(\mathcal{X}))$ and resampling operator $R_n : \mathcal{P}(\mathcal{P}(\mathcal{X})) \to \mathcal{P}(\mathcal{P}(\mathcal{X}))$ from Theorem A.2. The standard Monte Carlo set for p is just $R_n\mathbf{p}$. To show that $R_n\mathbf{p} \to \mathbf{p}$, weakly in $\mathcal{P}(\mathcal{P}(\mathcal{X}))$, as

$n \to \infty$, we must show that for every continuous, bounded, real-valued function $F : \mathcal{P}(\mathcal{X}) \to \mathbb{R}$,

$$\lim_{n\to\infty} (R_n \mathbf{p})\, F \to \mathbf{p}\, F = F(p(\mathbf{x})).$$

Now it's a matter of stringing together definitions:

$$\begin{aligned}
(R_n\mathbf{p})F &= \textstyle\int_{\mu\in\mathcal{P}(\mathcal{X})} F(\mu)(R_n\,\mathbf{p})(d\mu) && \text{by equation 2.7}\\
&= \textstyle\int_{\mu\in\mathcal{P}(\mathcal{X})} F(\mu)(r_n(p(\mathbf{x})))(d\mu) && \text{by definition of } R_n \text{ (A.5)}\\
&= F_{r_n}(p(\mathbf{x})) && \text{by equation A.4}\\
&\to F(p(\mathbf{x})) && \text{by Lemma A.1,}
\end{aligned}$$

which completes the proof. ■

Finally, we can use the resampling operator R_n to prove that the application of dynamics has the expected effect on particle sets.

Theorem A.4 *(This is the same as Theorem 2.3, page 20.) Let S, S' be particle sets, $p(\mathbf{x})$ a probability distribution on the configuration space $\mathcal{X}$, and $p(\mathbf{x}'|\mathbf{x})$ a conditional distribution. Suppose S represents $p(\mathbf{x})$ and S' is obtained from S by applying the dynamics $p(\mathbf{x}'|\mathbf{x})$. In addition, suppose S was produced from some other particle set by a resampling operation. Then S' represents the distribution $p'(\mathbf{x}')$ given by*

$$p'(\mathbf{x}') = \int_{\mathbf{x}} p(\mathbf{x}'|\mathbf{x})p(\mathbf{x})\, d\mathbf{x}.$$

Proof. In the now familiar style, we abstract the application of dynamics to an operator on $\mathcal{P}(\mathcal{P}(\mathcal{X}))$. Let w be the operator on $\mathcal{P}(\mathcal{X})$ which takes $\mu(\mathbf{x}) \in \mathcal{P}(\mathcal{X})$ to $\int p(\mathbf{x}'|\mathbf{x})d\mu(\mathbf{x})$, and let its "uncle" operator W on $\mathcal{P}(\mathcal{P}(\mathcal{X}))$ be defined for $\Phi \in \mathcal{P}(\mathcal{P}(\mathcal{X}))$ and bounded continuous $F : \mathcal{P}(\mathcal{X}) \to \mathbb{R}$ by

$$(W\Phi)F = \int F(w\mu)\Phi(d\mu).$$

It is easy to show W is a continuous operator, and that therefore $WS_n \to W\mathbf{p} = \mathbf{p}'$. Unfortunately, without the special resampling assumption in the theorem, it is not true that $S'_n = WS_n$; for this to be true, our recipe for applying dynamics would have to replace equation 2.13 by

$$\begin{aligned}
&\mathbf{x}_i^{(n)'} \vdash p(\mathbf{x}_i^{(n)'}|\mathbf{x}_j^{(n)}) \text{ with probability } \pi_j^{(n)}\\
&\pi_i^{(n)'} = 1/n. \qquad\qquad \text{(A.6)}
\end{aligned}$$

(See page 13 for the definition of $\vdash$.) However, take the special case that S *is* the result of resampling some other particle set, say U. So if R_n is the resampling operator defined in the proof of Theorem 2.4, we have $S_n = R_nU$. Then it so happens that applying dynamics via equation 2.13 to S_n has precisely the same effect as applying dynamics via equation A.6 directly to U_n (just check the definitions — imagine the actual algorithm you would follow to implement each, and you will find they are the same). So we get $S'_n = WU_n \to W\mathbf{p} = \mathbf{p}'$ as desired. ∎

Remark. This kind of limbo-dancing begs the question: why not define the application of dynamics using equation A.6 rather than equation 2.13? The answer is that it is cleaner — especially when trying to come up with efficient algorithms like partitioned sampling (Chapter 7) — to be able to think of dynamics as affecting only the positions of particles, not their weights.

B

Appendix

B.1 Summary of notation

$\mathcal{X}$	finite-dimensional shape space in which contour may deform, assumed to be compact (page 4)
$\mathbf{x}$	contour configuration, $\in \mathcal{X}$ (page 5)
$\vdash$	draw randomly from a distribution, independently of other draws from this distribution (page 13)
$\mathcal{P}(\mathcal{X})$	set of all probability measures on $\mathcal{X}$ (page 14)
$\mathcal{P}(\mathcal{P}(\mathcal{X}))$	set of all probability measures on $\mathcal{P}(\mathcal{X})$ (page 14)
$p, q, \ldots$	elements of $\mathcal{P}(\mathcal{X})$ (page 15)
$\mathbf{p}, \mathbf{q}, \ldots$	corresponding elements of $\mathcal{P}(\mathcal{P}(\mathcal{X}))$, e.g. $\mathbf{p}$ assigns mass 1 to p and 0 to all other distributions (page 15)
$\sim$	(on Condensation diagram) perform the resampling operation (page 10)
$\sim f$	(on Condensation diagram) perform the weighted resampling operation with respect to importance function f (page 130)
$\times f$	(on Condensation diagram) multiply by f (page 10)
$* f$	(on Condensation diagram) apply dynamics f (page 10)

$\mathcal{G}(z\|\nu)$	model error distribution (page 39)
q_{ij}	non-detection probability: probability of detecting i features on a line with intersection number j (page 40)
$b(n), b_L(n)$	background distribution: probability of detecting n features on a measurement line of length L lying in background clutter (page 40)
$c_{\mathbf{x}}(i), c(i), c$	intersection number of the ith measurement line with configuration $\mathbf{x}$ (page 42)
$\boldsymbol{\nu}_{\mathbf{x}}(i), \boldsymbol{\nu}(i), \boldsymbol{\nu}$	intersection innovations of the ith measurement line with configuration $\mathbf{x}$ (page 42)
$p_c(n; \mathbf{z}), p_c(n; \mathbf{z}\|\boldsymbol{\nu})$	probability density function for generative model on single measurement line with intersection number c and intersection innovation $\boldsymbol{\nu}$ (page 42)

Bibliography

Adelson, E. and Bergen, J. (1991). The plenoptic function and the elements of early vision. In *Computational Models of Visual Processing*, 3–20. MIT Press.

Akaike, H. (1974). A new look at the statistical model identification. *IEEE Transactions on Automatic Control*, 19, 6, 716–723.

Amir, A. and Lindenbaum, M. (1996). Grouping based non-additive verification. Technical Report 9518, Center for Intelligent Systems, Technion.

Astrom, K. and Wittenmark, B. (1984). *Computer Controlled Systems.* Addison Wesley.

Baker, S., Szeliski, R., and Anandan, P. (1998). A layered approach to stereo reconstruction. In *Proc. Conf. Computer Vision and Pattern Recognition*, 434–441.

Bar-Shalom, Y. and Fortmann, T. (1988). *Tracking and Data Association.* Academic Press.

Bartels, R., Beatty, J., and Barsky, B. (1987). *An Introduction to Splines for use in Computer Graphics and Geometric Modeling.* Morgan Kaufmann.

Baumberg, A. and Hogg, D. (1994). Learning flexible models from image sequences. In *Proc. 3rd European Conf. Computer Vision*, 299–308. Springer-Verlag.

Black, M. and Fleet, D. (1999). Probabilistic detection and tracking of motion discontinuities. In *Proc. 7th Int. Conf. on Computer Vision*, 551–558.

Black, M. and Jepson, A. (1998). A probabilistic framework for matching temporal trajectories: Condensation-based recognition of gestures and expressions. In *Proc. 5th European Conf. Computer Vision*, 1, 909–924, Freiburg, Germany. Springer Verlag.

Black, M. and Yacoob, Y. (1995). Tracking and recognizing rigid and non-rigid facial motions using local parametric models of image motion. In *Proc. 5th Int. Conf. on Computer Vision*, 374–381.

Blake, A., Bascle, B., Isard, M., and MacCormick, J. (1998). Statistical models of visual shape and motion. *Phil. Trans. R. Soc. A.*, 356, 1283–1302.

Blake, A., Curwen, R., and Zisserman, A. (1993). A framework for spatio-temporal control in the tracking of visual contours. *Int. J. Computer Vision*, 11, 2, 127–145.

Blake, A. and Isard, M. (1998). *Active Contours.* Springer-Verlag.

Blake, A., Isard, M., and Reynard, D. (1995). Learning to track the visual motion of contours. *J. Artificial Intelligence*, 78, 101–134.

Blake, A. and Yuille, A., editors (1992). *Active Vision.* MIT.

Bobick, A. and Wilson, A. (1995). A state-based technique for the summarisation and recognition of gesture. In *Proc. 5th Int. Conf. on Computer Vision*, 382–388.

Boden, M. (1990). *The Philosophy of Artificial Intelligence.* Oxford University Press.

Bozic, S. (1979). *Digital and Kalman Filtering.* Arnold.

Bregler, C. and Malik, J. (1998). Tracking people with twists and exponential maps. In *Proc. Conf. Computer Vision and Pattern Recognition*, 8–15.

Cameron, A. and Durrant-Whyte, H. (1988). Optimal sensor placement. Technical Report OURRG-99-10, Oxford University Robotics Research Group.

Carpenter, J., Clifford, P., and Fearnhead, P. (1997). An improved particle filter for non-linear problems. Technical report, Dept. of Statistics, University of Oxford. Available from `www.stats.ox.ac.uk/~clifford/index.html`.

Carpenter, J., Clifford, P., and Fearnhead, P. (1999). Building robust simulation-based filters for evolving data sets. Technical report, Dept. of Statistics, University of Oxford. Available from `www.stats.ox.ac.uk/~clifford/index.html`.

Chandler, D. (1987). *Introduction to Statistical Mechanics.* Oxford University Press.

Chellappa, R. and Chatterjee, S. (1985). Classification of textures using Gaussian Markov random fields. *IEEE Trans. Acoust., Speech, Signal Processing*, ASSP-33, 959–963.

Cootes, T., Page, G., Jackson, C., and Taylor, C. (1996). Statistical grey-level models for object location and identification. *J. Image and Vision Computing*, 14, 533–540.

Cootes, T. and Taylor, C. (1992). Active shape models. In *Proc. British Machine Vision Conf.*, 265–275.

Cootes, T. and Taylor, C. (1996). Locating objects of varying shape using statistical feature detectors. In *Proc. 4th European Conf. Computer Vision*, 465–474.

Cootes, T., Taylor, C., Cooper, D., and Graham, J. (1995). Active shape models — their training and application. *Computer Vision and Image Understanding*, 61, 1, 38–59.

Cox, I. and Hingorani, S. (1996). An efficient implementation of Reid's multiple hypothesis tracking algorithm and its evaluation for the purpose of visual tracking. *IEEE Trans. on Pattern Analysis and Machine Intelligence*, 18, 2, 138–150.

Crisan, D., Gaines, J., and Lyons, T. (1998). Convergence of a branching particle method to the solution of the Zakai equation. *SIAM Journal on Applied Mathematics*, 58, 5, 1568–1590.

Crisan, D. and Lyons, T. (1997). Nonlinear filtering and measure-valued processes. *Probability Theory and Related Fields*, 109, 2, 217–244.

Crisan, D. and Lyons, T. (1999). A particle approximation of the solution of the Kushner-Stratonovitch equation. *Probability Theory and Related Fields*, 115, 4, 549–578.

Curwen, R. and Blake, A. (1992). Dynamic contours: real-time active splines. In Blake, A. and Yuille, A., editors, *Active Vision*, 39–58. MIT.

De Souza, K. M. A., Kent, J. T., and Mardia, K. V. (1997). Bayesian analysis of highly variable images. In *The Art and Science of Bayesian Image Analysis*. 17th Leeds Annual Statistical Research Workshop, Leeds.

Del Moral, P. (1998). Measure-valued processes and interacting particle systems: application to nonlinear filtering problems. *The Annals of Applied Probability*, 8, 2, 438–495.

Deutscher, J., Blake, A., North, B., and Bascle, B. (1999). Tracking through singularities and discontinuities by random sampling. In *Proc. 7th Int. Conf. on Computer Vision*, 1144–1150.

Doucet, A. (1998). On sequential simulation-based methods for Bayesian filtering. Technical Report CUED/F-INFENG/TR310, Dept. of Engineering, University of Cambridge. Available from `www.stats.bris.ac.uk:81/MCMC/pages/list.html`.

Duda, R. and Hart, P. (1973). *Pattern Classification and Scene Analysis*. John Wiley and Sons.

Eliot, T. S. (1942). *Little Gidding*. Faber.

Fleck, M., Forsyth, D., and Bregler, C. (1996). Finding naked people. In *Proc. 4th European Conf. Computer Vision*, 593–602, Cambridge, England.

Fu, K. (1982). *Syntactic Pattern Recognition.* Prentice-Hall.

Gelb, A., editor (1974). *Applied Optimal Estimation.* MIT Press, Cambridge, MA.

Geman, D., Geman, S., Graffigne, C., and Dong, P. (1990). Boundary detection by constrained optimisation. *IEEE Trans. on Pattern Analysis and Machine Intelligence*, 12, 7, 609–628.

Geman, S. and Geman, D. (1984). Stochastic relaxation, Gibbs distributions, and the Bayesian restoration of images. *IEEE Trans. on Pattern Analysis and Machine Intelligence*, 6, 6, 721–741.

Geweke, J. (1989). Bayesian inference in econometric models using Monte Carlo integration. *Econometrica*, 57, 1317–1339.

Gidas, B. (1989). A renormalisation group approach to image processing problems. *IEEE Trans. on Pattern Analysis and Machine Intelligence*, 11, 2, 164–180.

Gidas, B. and Mumford, D. (1999). Stochastic models for generic images. Available from `www.dam.brown.edu/people/mumford/`.

Gilks, W., Richardson, S., and Spiegelhalter, D. (1996). *Markov Chain Monte Carlo in Practice.* Chapman and Hall.

Gordon, N. (1997). A hybrid bootstrap filter for target tracking in clutter. *IEEE Trans. Aero. Elec. Systems*, 33, 353–358.

Gordon, N., Salmond, D., and Smith, A. (1993). Novel approach to nonlinear/non-Gaussian Bayesian state estimation. *IEE Proc. F, Radar and Signal Processing*, 140, 2, 107–113.

Green, P. (1995). Reversible jump Markov chain Monte Carlo computation and Bayesian model determination. *Biometrika*, 82, 4, 711–32.

Grenander, U. (1976–1981). *Lectures in Pattern Theory I, II and III.* Springer.

Grenander, U., Chow, Y., and Keenan, D. (1991a). *HANDS. A Pattern Theoretic Study of Biological Shapes.* Springer-Verlag. New York.

Grenander, U., Chow, Y., and Keenan, D. (1991b). *HANDS. A Pattern Theoretical Study of Biological Shapes.* Springer-Verlag. New York.

Grimson, W. (1990). *Object Recognition by Computer.* MIT Press.

Grimson, W. and Huttenlocher, D. (1990). On the verification of hypothesised matches in model-based recognition. In *Proc. 1st European Conf. Computer Vision*, 489–498.

Grimson, W., Huttenlocher, D., and Jacobs, D. (1992). A study of affine matching with bounded sensor error. In *Proc. 2nd European Conf. Computer Vision*, 291–306.

Hallinan, P., Gordon, G., Yuille, A., Giblin, P., and Mumford, D. (1999). *Two- and Three-Dimensional Patterns of the Face.* A. K. Peters.

Handschin, J. (1970). Monte Carlo techniques for prediction and filtering of non-linear stochastic processes. *Automatica*, 6, 555–563.

Handschin, J. and Mayne, D. (1969). Monte Carlo techniques to estimate the conditional expectation in multi-stage non-linear filtering. *International Journal of Control*, 9, 5, 547–559.

Hardy, G. H. (1940). *A Mathematician's Apology.* Cambridge University Press.

Haritaoglu, I., Harwood, D., and Davis, L. (1998). w^4s: a real-time system for detecting and tracking people in 2.5D. In *Proc. 5th European Conf. Computer Vision*, 1, 877–892, Freiburg, Germany. Springer-Verlag.

Heap, T. and Hogg, D. (1998). Wormholes in shape-space: Tracking through discontinuous changes in shape. In *Proc. 6th Int. Conf. on Computer Vision.*

Hinton, G., Williams, C., and Revow, M. (1992). Adaptive elastic models for hand-printed character recognition. *Advances in Neural Information Processing Systems*, 4, 512–519.

Intille, S., Davis, J., and Bobick, A. (1997). Real-time closed-world tracking. In *Proc. Conf. Computer Vision and Pattern Recognition*, 697–703.

Irani, M. and Anandan, P. (1998). A unified approach to moving object detection in 2D and 3D scenes. *IEEE Trans. on Pattern Analysis and Machine Intelligence*, 20, 6, 577–589.

Isard, M. (1998). *Visual motion analysis by probabilistic propagation of conditional density.* PhD thesis, University of Oxford.

Isard, M. and Blake, A. (1996). Visual tracking by stochastic propagation of conditional density. In *Proc. 4th European Conf. Computer Vision*, 343–356, Cambridge, England.

Isard, M. and Blake, A. (1998a). Condensation — conditional density propagation for visual tracking. *Int. J. Computer Vision*, 28, 1, 5–28.

Isard, M. and Blake, A. (1998b). Icondensation: unifying low-level and high-level tracking in a stochastic framework. In *Proc. 5th European Conf. Computer Vision*, 893–908.

Isard, M. and Blake, A. (1998c). A mixed-state Condensation tracker with automatic model switching. In *Proc. 6th Int. Conf. on Computer Vision*, 107–112.

Isard, M. and MacCormick, J. (2000). Hand tracking for vision-based drawing. Technical report, Visual Dynamics Group, Dept. Eng. Science, University of Oxford. Available from `www.robots.ox.ac.uk/~vdg`.

Jollike, I. (1986.). *Principal Component Analysis*. Springer-Verlag.

Jones, M. and Rehg, J. (1998). Statistical color models with applications to skin detection. In *Proc. Conf. Computer Vision and Pattern Recognition*. Also available from `www.crl.research.digital.com`.

Karatzas, I. and Shreve, S. (1988). *Brownian Motion and Stochastic Calculus*. Springer-Verlag.

Kass, M., Witkin, A., and Terzopoulos, D. (1987). Snakes: active contour models. In *Proc. 1st Int. Conf. on Computer Vision*, 259–268.

Kaucic, R. (1997). *Lip tracking for audio-visual speech recognition*. PhD thesis, University of Oxford.

Kent, J., Mardia, K., and Walder, A. (1996). Conditional cyclic Markov random fields. *Adv. Appl. Prob. (SGSA)*, 28, 1–12.

Kitagawa, G. (1996). Monte Carlo filter and smoother for non-Gaussian nonlinear state space models. *Journal of Computational and Graphical Statistics*, 5, 1, 1–25.

Kjeldsen, R. and Kender, J. (1996). Toward the use of gesture in traditional user interfaces. In *Proc. 2nd Int. Conf. on Automatic Face and Gesture Recognition*, 151–156.

Koller, D., Weber, J., and Malik, J. (1994). Robust multiple car tracking with occlusion reasoning. In *Proc. 3rd European Conf. Computer Vision*, 189–196. Springer-Verlag.

Kong, A., Liu, S., and Wong, W. (1994). Sequential imputations and Bayesian missing data problems. *J. Am. Stat. Assoc.*, 89, 425, 278–288.

Langevin, P. (1908). Sur la théorie du mouvement brownien. *C. R. Acad. Sci. Paris*, 146, 530–533.

Leung, T., Burl, M., and Perona, P. (1995). Finding faces in cluttered scenes using random graph matching. In *Proc. IEEE PAMI Conf.*, 637–644, Cambridge.

Liggett, T. (1985). *Interacting Particle Systems*. Springer-Verlag.

Liu, J, and Chen, R. (1995). Blind deconvolution via sequential imputations. *J. Am. Stat. Assoc.*, 90, 430, 567–576.

Liu, J. and Chen, R. (1998). Sequential Monte Carlo methods for dynamic systems. *J. Amer. Statist. Assoc.*, 93, 1032–1043.

Lowe, D. (1987a). Three-dimensional object recognition from single two-dimensional images. *J. Artificial Intelligence*, 31, 355–395.

Lowe, D. (1987b). The viewpoint consistency constraint. *Int. J. Computer Vision*, 1, 57–72.

Lowe, D. (1991). Fitting parameterised 3D models to images. *IEEE Trans. on Pattern Analysis and Machine Intelligence*, 13, 5, 441–450.

Lowe, D. (1992). Robust model-based motion tracking through the integration of search and estimation. *Int. J. Computer Vision*, 8, 2, 113–122.

Lutkepohl, H. (1993). *Introduction to Multiple Time Series Analysis*. Springer-Verlag, 2nd edition.

MacCormick, J. and Blake, A. (1998a). A probabilistic contour discriminant for object localisation. In *Proc. 6th Int. Conf. on Computer Vision*, 390–395.

MacCormick, J. and Blake, A. (1998b). Spatial dependence in the observation of visual contours. In *Proc. 5th European Conf. Computer Vision*, 765–781.

MacCormick, J. and Blake, A. (1999). A probabilistic exclusion principle for tracking multiple objects. In *Proc. 7th Int. Conf. on Computer Vision*, 1, 572–578.

MacCormick, J. and Isard, M. (2000). Partitioned sampling, articulated objects, and interface-quality hand tracking. In *European Conf. Computer Vision*.

Mackay, D. (1999). Introduction to Monte Carlo methods. In Jordan, M., editor, *Learning in Graphical Models*. MIT Press.

Malik, J., Belongie, S., Shi, J., and Leung, T. (1999). Textons, contours and regions: cue integration in image segmentation. In *Proc. 7th Int. Conf. on Computer Vision*, 918–925.

McKean, H. (1967). Propagation of chaos for a class of nonlinear parabolic equations. In Aziz, A., editor, *Lecture Series in Differential Equations (7)*, 177–194. Catholic Univ., Washington D.C.

Metropolis, N., Rosenbluth, A., Rosenbluth, M., Teller, A., and Teller, E. (1953). Equation of state calculations by fast computing machines. *Journal of Chemical Physics*, 6, 1087–1092.

Miller, M., Srivastava, A., and Grenander, U. (1995). Conditional-mean estimation via jump-diffusion processes in multiple target tracking/recognition. *IEEE Transactions on Signal Processing*, 43, 11, 2678–2690.

Mumford, D. (1993). Elastica and computer vision. In *Algebraic Geometry and its Applications*, 507–518. Springer-Verlag.

Mumford, D. and Gidas, B. (1999). Stochastic models for generic images. Available from `www.dam.brown.edu/people/mumford/`.

Mumford, D. and Nitzberg, M. (1990). The 2.1D sketch. In *Proc. 3rd International Conf. Computer Vision*, 138–144.

Mumford, D. and Shah, J. (1985). Boundary detection by minimising functionals. In *Proc. Conf. Computer Vision and Pattern Recognition*, 22–26.

Mumford, D. and Shah, J. (1989). Optimal approximation by piecewise smooth functions. *Comm. Pure and Appl. Math.*, 42, 577–685.

Neal, R. (1998). Annealed importance sampling. Technical Report 9805, Dept. of Statistics, University of Toronto.

North, B. (1998). *Learning dynamical models for visual tracking.* PhD thesis, Department of Engineering Science, Oxford University.

Papoulis, A. (1990). *Probability and Statistics.* Prentice-Hall.

Paragios, N. and Deriche, R. (1998). A PDE-based level-set approach for detection and tracking of moving objects. In *Proc. 6th International Conf. Computer Vision*, 1139–45.

Pavlovic, V., Rehg, J., Cham, T., and Murphy, K. (1999). A dymanic Bayesian network approach to figure tracking using learned dynamic models. *Proc. 7th Int. Conf. on Computer Vision.*

Perez, P. and Heitz, F. (1996). Restriction of a Markov random field on a graph and multiresolution statistical image modelling. *IEEE Trans. Information Theory*, 42, 1, 180–190.

Peterfreund, N. (1998). Robust tracking with spatio-velocity snakes: Kalman filtering approach. In *Proc. 6th International Conf. Computer Vision*, 433–39.

Pitt, M. and Shepherd, N. (1997). Filtering via simulation and auxiliary particle filters. Technical report, Nuffield College, University of Oxford.

Propp, J. and Wilson, D. (1996). Exact sampling with coupled Markov chains and applications in statistical mechanics. *Random Structures and Algorithms*, 9, 223–252.

Rabiner, L. and Bing-Hwang, J. (1993). *Fundamentals of Speech Recognition.* Prentice-Hall.

Rasmussen, C. and Hager, G. (1998). Joint probabilistic techniques for tracking multi-part objects. In *Proc. Conf. Computer Vision and Pattern Recognition*, 16–21.

Reid, D. (1979). An algorithm for tracking multiple targets. *IEEE Transactions on Automatic Control*, 24, 6, 843–854.

Reynard, D., Wildenberg, A., Blake, A., and Marchant, J. (1996). Learning dynamics of complex motions from image sequences. In *Proc. 4th European Conf. Computer Vision*, 357–368, Cambridge, England.

Ripley, B. (1987). *Stochastic Simulation.* New York: Wiley.

Ripley, B. and Sutherland, A. (1990). Finding spiral structures in images of galaxies. *Phil. Trans. R. Soc. Lond. A.*, 332, 1627, 477–485.

Robertson, R. and Schwertassek, R. (1988). *Dynamics of Multibody Systems.* Springer-Verlag.

Rothwell, C. (1996). Reasoning about occlusions during hypothesis verification. In *Proc. 4th European Conf. Computer Vision*, 599–609.

Rothwell, C., Zisserman, A., Mundy, J., and Forsyth, D. (1992). Efficient model library access by projectively invariant indexing functions. In *Proc. Conf. Computer Vision and Pattern Recognition*, 109–114.

Rubin, D. (1988). Using the SIR algorithm to simulate posterior distributions. In *Bayesian Statistics*, 3, 395–402. Oxford University Press.

Rue, H. and Husby, O. (1997). Identification of partly destroyed objects using deformable templates. Technical Report 1/97, NTNU Statistics. available from `www.math.ntnu.no/preprint/statistics`.

Shao, Y. and Mayhew, J. (1996). A stochastic framework for object localisation. In *Proc. British Machine Vision Conf.*, 203–212.

Shimshoni, I. and Ponce, J. (1995). Probabilistic 3D object recognition. In *Proc. 5th Int. Conf. on Computer Vision*, 488–493.

Smith, A. and Gelfand, A. (1992). Bayesian statistics without tears: a sampling–resampling perspective. *The American Statistician*, 46, 84–88.

Soh, J., Chun, B., and Wang, M. (1994). An edge-based approach to moving object location. In *Image and Video Processing II*, 2182, 132–41. SPIE.

Stewart, L. (1983). Bayesian analysis using Monte Carlo integration — a powerful methodology for handling some difficult problems. *The Stat.*, 32, 195–200.

Stockman, G., Kopstein, S., and Benett, S. (1982). Matching images to models for registration and object detection via clustering. *IEEE Trans. on Pattern Analysis and Machine Intelligence*, 3, 3, 229–241.

Sullivan, J., Blake, A., Isard, M., and MacCormick, J. (1999). Object localisation by Bayesian correlation. In *Proc. 7th Int. Conf. on Computer Vision*, 1068–1075.

Terzopoulos, D. and Szeliski, R. (1992). Tracking with Kalman snakes. In Blake, A. and Yuille, A., editors, *Active Vision*, 3–20. MIT.

Uhlenbeck, G. and Ornstein, L. (1930). On the theory of Brownian motion. *Physical Review*, 36, 823–841.

Wang, J. and Adelson, E. (1993). Layered representation for motion analysis. In *Proc. Conf. Computer Vision and Pattern Recognition*, 361–366.

Wildenberg, A. (1997). *Learning and initialisation for visual tracking*. PhD thesis, University of Oxford.

Williams, L. and Thornber, K. (1999). A comparison of measures for detecting natural shapes in cluttered backgrounds. In *Proc. 5th European Conf. Computer Vision*, 432–448.

Winkler, G. (1995). *Image Analysis, Random Fields and Dynamic Monte Carlo Methods*. Springer-Verlag.

Wu, Y. N., Zhu, S. C., and Liu, X. (1999). Equivalence of Julesz and Gibbs texture ensembles. In *Proc. 7th Int. Conf. on Computer Vision*, 1025–32.

Zhu, S., Wu, Y., and Mumford, D. (1997). Minimax entropy principle and its application to texture modelling. *Neural Computation*, 9, 1627–60.

Zhu, S., Wu, Y., and Mumford, D. (1998). Filters, random fields and maximum entropy (FRAME). *Int. J. Computer Vision*, 27, 2, 107–126.

Index

MIX
Papier aus verantwortungsvollen Quellen
Paper from responsible sources
FSC® C105338

If you have any concerns about our products,
you can contact us on
ProductSafety@springernature.com

In case Publisher is established outside the EU,
the EU authorized representative is:
Springer Nature Customer Service Center GmbH
Europaplatz 3, 69115 Heidelberg, Germany

Printed by Libri Plureos GmbH
in Hamburg, Germany